The Natural Philosophy of James Clerk Maxwell

This book provides an introductory yet comprehensive account of James Clerk Maxwell's (1831–1879) physics and worldview. The argument is structured by a focus on the fundamental themes which shaped Maxwell's science: analogy and geometry, models and mechanical explanation, statistical representation and the limitations of dynamical reasoning, and the relation between physical theory and its mathematical description. This approach, which considers his physics as a whole, bridges the disjunction between Maxwell's greatest contributions: the concept of the electromagnetic field and the kinetic theory of gases. Maxwell's work and ideas are viewed historically in terms of his indebtedness to the scientific and cultural traditions, of Edinburgh experimental physics and of Cambridge mathematics and philosophy of science, that nurtured his career.

PETER HARMAN is Professor of the History of Science at Lancaster University. He is the editor of *The Scientific Letters and Papers of James Clerk Maxwell* (also published by Cambridge University Press). The present book derives from his lectures as Zeeman Visiting Professor of the History of Physics at the University of Amsterdam.

Reviews from the hardback edition

'By stressing the deep traditionalism of Maxwell's enterprise, Harman makes him a participant in a philosophical conversation about matter, motion and mind, well-established by the time of Newton and continuing into this century.'
Simon Schaffer, *London Review of Books*

'The sections on Maxwell's physics in this book are presented with admirable clarity, particularly the intricacies of how his theory of electromagnetism developed, which are usually difficult to follow. This presentation gives the general reader a clear overview of Maxwell's work, while the discussion of how Maxwell's physics developed with time provides a good insight into how philosophical ideas matured.' Elizabeth Garber, *Physics World*

'. . . not the least of his talents is his ability to present brief, informed, and intelligible accounts of the results of [technical] analysis and use them in his argumentation.' Ole Knudsen, *Centaurus*

'The book is wonderfully informative and insightful throughout; its merit lies as much in stimulating questions as in providing answers.'
Daniel Siegal, *American Scientist*

'"Natural Philosopher" probably captures as much of Maxwell as any single label can, and Harman is to be commended for bringing a measure of synthesis to our picture of one of the most important but elusive figures of Victorian science.' Bruce J. Hunt, *Isis*

The Natural Philosophy of
James Clerk Maxwell

P. M. HARMAN

CAMBRIDGE UNIVERSITY PRESS
Cambridge, New York, Melbourne, Madrid, Cape Town, Singapore, São Paulo

Cambridge University Press
The Edinburgh Building, Cambridge CB2 2RU, UK

Published in the United States of America by Cambridge University Press, New York

www.cambridge.org
Information on this title: www.cambridge.org/9780521561020

First published 1998
First paperback edition 2001

A catalogue record for this publication is available from the British Library

ISBN-13 978-0-521-56102-0 hardback
ISBN-10 0-521-56102-7 hardback

ISBN-13 978-0-521-00585-2 paperback
ISBN-10 0-521-00585-X paperback

Transferred to digital printing 2005

For my mother and in memory of my father

Contents

Contents

Preface

This book is based on lectures I delivered in spring 1995 at the University of Amsterdam, as Zeeman Professor of the History of Physics. I am very grateful to the Pieter Zeeman Foundation, and especially to Anne Kox, for this generous invitation. On revising the text for publication I have retained traces of its original presentation as a series of lectures, as an aid to clarity of exposition. I have aimed to provide an introductory yet comprehensive account of Maxwell's science and worldview, intended to be accessible to readers without specialised knowledge bearing on its subject.

The traditional term 'natural philosopher' may be aptly applied to a scientist who was also a scholar, deeply conscious of the historical roots and philosophical import of his physics. The chapters of this study are structured by the broad categories which shape Maxwell's natural philosophy: analogy and geometry, mechanical and statistical representation. Maxwell's work and ideas are located historically in terms of his indebtedness to the cultural and scientific traditions, of Edinburgh experimental physics and metaphysics, and of Cambridge mathematics and philosophy of science, that nurtured his career and intellectual development.

Historians of physics have traditionally described Maxwell's science by writing from the perspective of one or other of the two main fields of physics which he revolutionised: the theory of electromagnetism ('Maxwell's equations') and the electromagnetic theory of light, and the kinetic theory of gases and statistical physics. Some excellent work has been accomplished within this constraint, and this frame of reference is appropriate if the concern is with the development of 'Maxwell's equations' or his statistical methods in mathematical physics. But there has been little attempt to consider Maxwell's physics as a whole intellectual endeavour. In defining the historical Maxwell in terms of topics in physics, these studies have been based on a very restricted range of sources, and this has led to a limited reading of his intentions.

In seeking to provide a more comprehensive interpretation of Maxwell's intellectual outlook and practice, the argument here is based on study of the whole range of his writings. This book focuses on the fundamental themes of Maxwell's natural philosophy. It is not a biography or an account of Maxwell's career, though the presentation is broadly chronological and attention is paid to the scientific and cultural context. Nor does it provide minute analysis of special points of physics, though the argument is, on occasion, technical.

The argument draws extensively upon the manuscript materials collected in my edition of *The Scientific Letters and Papers of James Clerk Maxwell*, and provides a supplementary commentary to that edition, giving a more general and systematic account of Maxwell's science than is possible in the introductions to its volumes. To avoid burdening this book with an excessive scholarly apparatus, wherever appropriate, reference is made to the extensive editorial annotations in *Letters and Papers*, but the citations in the text and endnotes provide necessary documentation.

To date, two of the three planned volumes of *Letters and Papers* have been published; archival references are given in the endnotes for manuscripts which will be printed in its final volume. For permission to include these documents and to reproduce photographs in the plates I am grateful to the Syndics of the Cambridge University Library, the President and Council of the Royal Society, and the Cavendish Laboratory, Cambridge.

I appreciate my continued association with Cambridge University Press, and thank Simon Mitton for his kind encouragement; and I owe a debt of gratitude to Susan Bowring for her meticulous work as copy-editor. This book is a product of research carried out in the preparation of *Letters and Papers*, and I thank the Council of the Royal Society for a succession of research grants which have supported this project. But it has its origin in work I began many years ago, and I gratefully recollect the generous encouragement I received then, especially from Ted McGuire. For granting me leave of absence to assist its completion I am grateful to colleagues in the Department of History at Lancaster University.

Abbreviations

The following abbreviations are used in the text for references to frequently cited works. Citations are by volume, page, and (where appropriate) section numbers.

LP
The Scientific Letters and Papers of James Clerk Maxwell, ed. P. M. Harman, 2 vols. to date (Cambridge, 1990, 1995)

SP
The Scientific Papers of James Clerk Maxwell, ed. W. D. Niven, 2 vols. (Cambridge, 1890)

Treatise
James Clerk Maxwell, *A Treatise on Electricity and Magnetism*, 2 vols. (Oxford, 1873)

Plates

1 Introduction: Maxwell and the history of physics

Reviewing James Clerk Maxwell's *Treatise on Electricity and Magnetism* on its publication in 1873, Peter Guthrie Tait described his friend as having 'a name which requires only the stamp of antiquity to raise it almost to the level of that of Newton'. At the time Tait's enthusiasm may have seemed to verge on hyperbole; yet such has been the judgement of posterity. Tait accurately highlighted as the cardinal features of the *Treatise* Maxwell's demonstration of 'the connection between radiation and electrical phenomena', and his achievement in having 'upset completely the notion of *action at a distance*'.[1]

Maxwell's theory of the electromagnetic 'field', expounded in the *Treatise*, supposes that electric and magnetic forces are mediated by the agency of the 'field', contiguous elements of the space in the neighbourhood of the electric or magnetic bodies, the 'field' being embodied by an ether. The impact of the *Treatise* was at first muted, and at the time Maxwell's reputation rested largely on his work on molecular physics and gases. But within a few years of Maxwell's death in 1879 his theory of the electromagnetic field shaped the work of 'Maxwellian' physicists (George Francis FitzGerald, Oliver Heaviside, Joseph John Thomson and others). Following Heinrich Hertz's production and detection of electromagnetic waves in 1888, Maxwell's field theory and electromagnetic theory of light was accepted, notably by the leading theorist of the 1890s Henrik Antoon Lorentz, and came to be regarded as one of the most fundamental of all physical theories. 'Maxwell's equations' were accorded the status of Newton's laws of motion; and the theory was basic to the new technology of electric power, telephony and radio.

Maxwell's field theory and molecular physics achieved pre-eminence in the 'classical' physics of the nineteenth century, and mark an epoch in the history of the science, establishing his special place in the history of physics alongside Isaac Newton and Albert Einstein. The revolution in the structure of physical theory which has occurred in the twentieth century has reinforced rather than qualified Maxwell's unique status. His contributions to fundamental physics – the theory of the physical field and the electromagnetic theory of light, and the description of the motions of gas molecules by a statistical function – stand as progenitors of the relativity and quantum theories.

1 Introduction: Maxwell and the history of physics

In his famous paper on the theory of light in 1905 Albert Einstein pointed to a 'profound formal distinction' between field theory, where continuous spatial functions specify the electromagnetic state of a space, and molecular theory, where the state of a body is specified by the positions and velocities of a finite number of particles.[2] Writing in 1931 on the centenary of Maxwell's birth, Einstein appealed to a 'programme which may suitably be called Maxwell's: the description of Physical Reality by fields which satisfy without singularity a set of partial differential equations', in support of his contention that classical field theory should serve as the starting-point from which quantum rules emerge.[3] Rendition of Maxwell's outlook in these terms was of course intended to evoke Einstein's own endeavours and aspirations, but underlines the gap (as he understood it) between the primacy of fields (Maxwell's electromagnetic theory) and quantum theory (the statistical physics of particles).

The historical literature has naturally placed special emphasis on Maxwell's canonical contributions to fundamental physics, his field theory and statistical physics. The representation of the conceptual structure of physics as a duality of fields and particles, of electromagnetism and statistical physics, seen as having its historical roots in Maxwell's work and its contemporary expression in general relativity and quantum theory, has fostered this focus on the twin glories of Maxwell's science. Viewing Maxwell's physics from the vantage point of fields and particles, the historical analysis of his science has been defined by two areas of physics, electromagnetism and the kinetic theory of gases.

Writing in 1856, at the outset of his career, Maxwell suggests that nature may not be analogous to a 'book', envisaged as an ordered unity, but that the appropriate metaphor is a 'magazine', implying a collection of disconnected parts and a disparity in theorising.

> Perhaps the 'book', as it has been called, of nature is regularly paged . . . but if it is not a 'book' at all, but a *magazine*, nothing is more foolish to suppose that one part can throw light on another.
>
> (*LP*, **1**: 382)

Maxwell's theory of the electromagnetic field and his statistical physics illustrate such a disjunction in physical science. But the customary rendition of his physics in terms of a duality of electromagnetism and the kinetic theory of gases has provided a too restricted basis for analysing the structure of his scientific worldview. To provide a more discriminating framework for historical analysis, responsive to the categories of his own evolving conceptualisations, I will discuss his physics under broad thematic headings, categories which transcend the customary duality – of field theory (electromagnetism) and molecular physics (gas theory) – which has been traditionally used to characterise his science.

This representation of the structure of physical theory, as a dualism of fields and particles, contrasts with the scope of physics as understood at the time Maxwell began his career, and attests to the enormous impact of Maxwellian physics in marking an epoch in the science. Between 1800 and 1850 the science of physics was developing into a recognisably modern form: the study of mechanics, optics, heat, electricity, and magnetism, employing a mathematical and experimental methodology.

Around 1850, when Maxwell began his career, thermodynamics was in its infancy, resting on the two newly established laws of thermodynamics (the law of the conservation of energy, and the directional flow of heat from hot to cold bodies), while only the first steps had been taken to impose a mathematical structure on Michael Faraday's innovations in the study of electricity and magnetism. In the 1850s the law of the conservation of energy, as a cardinal element of the mechanical worldview of particles of matter in motion, came to be seen as fundamental to physical explanation; it was basic to Maxwell's subsequent achievement. In 1854 William Thomson (later Lord Kelvin), at the time Maxwell's guide to current work in physics, declared that the statement of the energy principle was 'the greatest reform that physical science has experienced since the days of Newton'.[4] Around 1850 the science of physics came to be defined in terms of the unifying role of the concept of energy and the programme of mechanical explanation.[5] Quantification, the search for mathematical laws, and precision measurement, the attainment of accurate values in experimentation, came to be seen as normative in physical science.

Maxwell shaped physical theory into its Maxwellian form by building on the work of his immediate predecessors – Hermann Helmholtz and Thomson in energy physics, Faraday and Thomson in field theory, Thomson and Rudolf Clausius in thermodynamics, and Clausius in the theory of gases. Maxwell's great achievements – the unification of optics (the theory of the luminiferous ether) and electromagnetism in his electromagnetic theory of light, the application of particle mechanics to understand the properties of gases and the foundations of the science of thermodynamics – rested on understanding the analogies and unities between the disparate themes of contemporary physics.

He himself emphasised the value of the 'cross-fertilization of the sciences' (*SP*, **2**: 744), evoking the image of bees pollinating flowers; and from the outset, he stressed the creative value of grasping the 'physical analogies' between different phenomena. Fundamental to these analogies and unities was understanding the relation between the language of mathematics and the structure of physical reality, between mathematical abstraction and the data of physical experiment, the

> hidden and dimmer region where Thought weds Fact, where the mental operation of the mathematician and the physical action of the molecules are seen in their true relation. (*SP*, **2**: 216)

The mechanical or dynamical worldview, which dominated the programme of physical explanation in the nineteenth century, shaped Maxwell's scientific theorising. But his attitude to mechanical explanation was complex. There was a tension in his thought between physical and mathematical models of mechanical systems; and his reflections on the relationship between mechanical representations and physical reality shaped his evolving programme of explanation. His introduction of statistical reasoning in the theory of gas molecules, and discussion of the instability and unpredictability of mechanical systems, led him to qualify his commitment to mechanism. The role of mechanical principles in his physics is complex and variegated, and to provide a preliminary perspective on these issues I will outline some of the central elements of his physics.[6]

Writing to Thomson in February 1854, after graduating from Cambridge University, Maxwell declared his intention to attack the science of electricity. In the 'Preface' to his *Treatise on Electricity and Magnetism* (1873) he recalled that he had commenced his work by study of Michael Faraday's *Experimental Researches in Electricity* (1839–55). Faraday had explained magnetism in terms of lines of force traversing space, and electrostatics by the mediation of forces by the dielectric. In 1845, by drawing on the analogy between electrostatics and the conduction of heat, which opened up applications of potential theory, Thomson showed that Faraday's ideas were compatible with the mathematical theory of electrostatics based on direct action at a distance. Thomson went on to develop theorems which could be applied to Faraday's discoveries in magnetism.

Guided by Thomson, Maxwell advanced beyond the work of his mentor in grappling comprehensively with Faraday's concept of the 'magnetic field'. In his paper 'On Faraday's lines of force' (1856) he presented a 'geometrical model' of lines of force in space, a representation resting on potential theory and the geometry of orthogonal surfaces, given embodiment by the 'physical analogy' of the flow of an incompressible fluid (*SP*, **1**: 156–8). He formulated theorems of electromagnetism, expressing the relation between magnetic forces and electric currents.

The analogy of streamlines in a fluid was proposed as illustrative of the geometry of the field; but Maxwell sought a theory of the field grounded on the mechanics of a mediating ether. He found its basis in Thomson's proposal that the Faraday magneto-optical rotation could be explained by the rotation of vortices in an ether. Maxwell began to develop the idea of orienting molecular vortices along magnetic field lines, culminating in the publication of his paper 'On physical lines of force', published in four parts in 1861–2. His physical model of vortices and 'idle wheel' particles, an ether model which is the most famous image in nineteenth-century physics, provides mechanical correlates for electromagnetic quantities in his field equations. The angular velocity of the vortices corresponds to the magnetic field intensity, and the translational flow of the idle wheel particles to the flow of an electric current. But he emphasised that while the theory was 'mechanically conceivable', the model itself was hardly 'a mode of connexion existing in nature' (*SP*, **1**: 486).

During the summer of 1861, while modifying the ether model to encompass electrostatics, he obtained an unexpected consequence, the 'Electromagnetic Theory of Light', as he termed his theory in 1864 (*LP*, **1**: 194). He introduced a 'displacement' of electricity as an electromagnetic correlate of the elastic deformation of the vortices, an elastic property which allowed for the propagation of transverse shear waves. He established the close agreement between the velocity of propagation of waves in an electromagnetic medium (which he demonstrated to be given by the ratio of electrostatic and electromagnetic units, established experimentally), and the measured velocity of light. This led him to claim the unification of optics and electromagnetism: that light consists in the vibrations of the '*same medium which is the cause of electric and magnetic phenomena*' (*SP*, **1**: 500). He completed the theory by a quantitative account of the magneto-optic effect in terms of the rotation of molecular vortices.

He was, however, dissatisfied with the appeal to a mechanical model, and sought to base his theory on firmer theoretical ground and to confirm its experimental basis. In 1862 he joined the British Association committee on electrical standards; and in May and June 1863, with Fleeming Jenkin and Balfour Stewart, made an accurate measurement of electrical resistance in absolute units (of time, mass and space). As part of the committee's report in 1863, Maxwell and Jenkin wrote a paper introducing dimensional notation: for every electrical quantity there are two absolute units, the electrostatic and the electromagnetic, and the ratio of these units is a power of a constant with the dimensions of a velocity. As Maxwell had established, this ratio was the velocity of waves in an electromagnetic medium. In 1868 he obtained a new value for the ratio of units by an experiment balancing the (electrostatic) force between two oppositely charged discs against the (electromagnetic) repulsion between two current-carrying coils, providing support for his theory.

In 'A dynamical theory of the electromagnetic field' (1865) Maxwell achieved a more general and systematic presentation of his theory. The ether model was abandoned, yet he retained the mechanical foundations of his theory by grounding the eight sets of 'general equations of the electromagnetic field' (the forerunners of the four 'Maxwell equations', as reformulated in the 1880s by Heaviside and Hertz) on the Lagrangian formalism of abstract dynamics. But in detaching his theory from the model he altered the interpretation of the displacement current, leading to a loss of consistency, a problem resolved in the *Treatise* where he interprets the displacement current as manifested as electric charge emergent from the field.

In the *Treatise* Maxwell emphasises the expression of physical quantities free from direct representation by a mechanical model. He enlarges the physical geometry and mechanical foundations of his earlier papers, deploying four fundamental mathematical ideas: quaternions (vector concepts), integral theorems (Stokes' theorem), topological concepts, and the Lagrange–Hamilton method of analytical dynamics (as developed by Thomson and Tait in their *Treatise on Natural Philosophy* of 1867). Maxwell's distinctive theory becomes most explicit in the final part of the work, on electromagnetism: here he presents the general equations of the electromagnetic field, the electromagnetic theory of light, and the dynamical basis of his field theory. The work concludes with a rebuttal of contemporary theories deriving from the tradition of considering forces acting at a distance without the mediation of a 'field'. He argues that these theories cannot satisfactorily explain the transmission of energy, for 'there must be a medium or substance in which the energy exists'. Mediation by an ether, the seat of the electromagnetic field, was the keystone of his theory (*Treatise*, **2**: 438 (§866)).

Maxwell's gas theory also has its origins in work undertaken upon his graduation from Cambridge. In March 1855 the subject of the University's

Adams Prize for 1857 was advertised as a study of 'The Motions of Saturn's Rings'. On revising his prize-winning essay for publication, Maxwell concluded that the ring system consists of concentric rings of satellites; this formed the argument of his memoir *On the Stability of the Motion of Saturn's Rings* (1859). The problems generated by this investigation played a role in initiating his work on the kinetic theory of gases in 1859. In considering the rings as a system of particles he noted that he was unable to compute the trajectories of these particles 'with any distinctness' (*SP*, **1**, 354). The Saturn's rings problem alerted him to discuss the complex motions of gas particles, where he introduced a probabilistic argument.

On completing his work on Saturn's rings, Maxwell had drawn on data on gas viscosity to establish the effect of friction in disturbing the stability of the rings. Alerted to gas viscosity and particle collisions, in spring 1859 he became interested in a paper by Rudolf Clausius on the theory of gases considered as particles in motion. To explain the slow diffusion of gas molecules, Clausius had calculated the probability of a molecule travelling a given distance (the mean free path) without collision. Maxwell had been interested in probability theory as early as 1850; and he advanced on Clausius' procedure by introducing a statistical formula for the distribution of velocities among gas molecules, a function identical in form to the distribution formula in the theory of errors. Beginning as an 'exercise in mechanics' (*LP*, **1**: 610), his work generated results in molecular physics. He was able to calculate the mean free path of molecules, and established the unexpected result that the viscosity of gases was independent of their density.

He turned to investigate the viscosity of gases at different temperatures and pressures, by observing the decay in the oscillation of discs torsionally suspended in a container, experiments presented as the Royal Society's Bakerian Lecture in 1866. He found that gas viscosity was a linear function of the absolute temperature; and he suggested, in his major paper 'On the dynamical theory of gases' (1867), that gas molecules should be considered as centres of force subject to an inverse fifth-power law of repulsion, a result in agreement with this experimental finding. He presented a new derivation of the distribution law, demonstrating that the velocity distribution would maintain a state of equilibrium unchanged by collisions. In drafting this paper he found that his theory seemed to have the consequence that energy could be abstracted from a cooling gas, a result in conflict with the second law of thermodynamics, stated in the early 1850s by Clausius and Thomson as denoting the tendency of heat to pass from warmer to colder bodies. While he corrected his argument and resolved the difficulty, it is likely that reflection on the problem led him to consider the bearing of his theory of gases on the interpretation of the second law of thermodynamics.

He first formulated the famous 'demon' paradox (the term was later coined by Thomson) in December 1867. By suggesting how a hot body could take heat from a colder one he showed that the second law of thermodynamics is a statistical regularity. Because of the statistical distribution of molecular velocities in a gas at equilibrium there will be spontaneous fluctuations of molecules taking heat from a cold body to a hotter one. But it would require the action of Maxwell's 'finite being', as he termed it (*LP*, **2**: 332), to manipulate molecules so as to produce an observable flow of heat from a cold body to a hotter one, and violate the second law of thermodynamics; hence the law is statistical and applies only to systems of molecules.

Maxwell amplified his argument to highlight a disjunction between the laws of mechanics and the second law of thermodynamics: this law is time-directional, expressing the irreversibility of physical processes, while the laws of mechanics are time-reversible. He maintained that the second law of thermodynamics is a statistical expression, not a dynamical theorem. In the 1870s, notably in a major paper on statistical mechanics written in 1878 (where he introduced the concept of ensemble averaging), he strove to clarify the relations between the dynamical and statistical descriptions of physical systems.

This cursory summary of Maxwell's most famous and enduring contributions to fundamental physics does not of course do justice to the physical and mathematical arguments upon which his theories rest. But this summary does indicate the centrality in his physics of issues such as the nature of physical analogies and of mechanical models, the relation between these models and more general dynamical principles, and the relation between dynamical laws and statistical explanations. These foundational issues transcend the disjunction between Maxwell's field and particle theories, between electromagnetism and the theory of gases.

The issue of Maxwell's commitment to mechanical explanation has loomed large in the critical and historical literature. In his famous 'Lectures on Molecular Dynamics and the Wave Theory of Light', delivered at The Johns Hopkins University in Baltimore in October 1884, William Thomson expressed strong criticism of Maxwell's approach to mechanical explanation. He professed 'immense admiration' for Maxwell's 'mechanical model of electromagnetic induction' – the ether model of 'On physical lines of force' – regarding the model as 'immensely instructive' and 'a step towards a definite mechanical theory of electro-magnetism'. But he bemoaned Maxwell's retreat from this approach to mechanical modelling. He considered Maxwell's electromagnetic theory of light, in the form presented in the *Treatise*, to be 'rather a backward step' in renouncing the provision of a mechanical model as the basis of a dynamical theory of electromagnetism. He contrasted Maxwell's theory, which he believed to be insufficiently grounded on mechanical principles, with the 'absolutely definite mechanical notion that is put before us by Fresnel and his followers', who had developed elastic solid theories of the luminiferous ether. He set out his own canon of mechanical explanation in a famous statement: 'as long as I cannot make a mechanical model all the way through I cannot understand; and that is why I cannot get the electromagnetic theory [of light]'.[7]

Reviewing the status of mechanical explanation some twenty years later, Pierre Duhem poured scorn on Thomson's claim for mechanical models as the sole basis for physical intelligibility. Duhem included in his critique 'that model of electrical actions which Maxwell built and for which Thomson has constantly professed his admiration',[8] the ether model of 'On physical lines of force'. Duhem has not been alone in dismissing the rationale of this ether model.[9] But discussion of the status of mechanism in Maxwell's physics has faltered because of imprecision in defining the problem.

Even within his theory of the electromagnetic field, where his mechanical outlook seems most emphatic, Maxwell's papers elaborate mechanical ideas in different senses. His paper 'On physical lines of force' (1861–2) is based on an explicitly mechanical outlook, yet he was flexible in his commitment to different elements of his ether model of vortex cells and 'idle wheel' particles. In his later writings, notably in the *Treatise*, he adopted an abstract approach to mechanical representation, based on the Lagrangian theory of a connected dynamical system, abandoning the attempt to provide a description of the internal characteristics of the electromagnetic field. But he still makes reference to formulating a 'complete dynamical theory' of the field in which 'the whole intermediate mechanism and details of the motion' would be studied (*Treatise*, **2**: 202 (§574)).

Within Maxwell's statistical theory of gases and thermodynamics the issue is even more difficult. He consistently presents his theory of gases as a 'dynamical' theory, by which he means a theory of particles in motion regulated by laws of forces. But he distinguishes the knowledge generated by the 'statistical method' of his theory of gases, which rests on probabilities, from the certain predictions of the 'dynamical method' which could trace and predict the trajectories of individual particles. The burden of the 'demon' paradox – that the second law of thermodynamics is a statistical theorem, not a dynamical law – underscored the limits of a physics based on purely dynamical principles. While he continued to stress the link between the dynamical and statistical specifications of a system, there is some divergence between his statistical method and his expression of dynamical theory.

Moreover, Maxwell questioned some of the basic assumptions of mechanism. In arguing that there are limitations in the explanatory power of dynamics, he pointed to the instability of a mechanical system at a point of singularity, where its trajectory could not be predicted. He drew the implication that while the universe was regulated by causal dynamical laws, these laws were not wholly deterministic. There are different strata in Maxwell's exposition and critique of dynamical principles; his arguments must be considered in their full variety and complexity. Maxwell's construal of mechanical explanation cannot be understood if the problem is viewed in narrow focus, limited to analysis of strands of his argument such as the role of mechanical models in electromagnetism.

It is the aim of this book to describe the structure of Maxwell's physical worldview, based on fields and statistical physics, and to elucidate its architectonic by tracing the motifs which thread their way through his natural philosophy. The term 'natural philosophy', which was becoming obsolete by Maxwell's time, seemingly at odds with the norms of the emerging community of 'physicists', is aptly descriptive of the ambition and scope of Maxwell's physics, which aimed, in traditional style, to lay the foundations of a scientific worldview. It was at the 1833 meeting of the recently formed British Association for the Advancement of Science that William Whewell, responding to the poet Coleridge's complaint that the term 'philosopher' was 'too wide and too lofty' for contemporary students of natural knowledge, proposed the term 'scientist'; this neologism served to demarcate 'science' (natural philosophy) from 'philosophy' (moral and metaphysical), and to emphasise the communality of the scientific enterprise. Whewell gave the term currency in his *Philosophy of the Inductive Sciences* (1840), where he also coined the neologism 'physicist' to describe the student of physics, investigating 'force, matter, and the properties of matter'.[10]

This shift in terminology matched a transformation within science itself:

the emergence in the nineteenth century of newly defined and specialised scientific disciplines with distinctive concepts and techniques for research. The transformation from an avocation to a vocational pursuit led to the emergence of a specialised and trained elite and to the proliferation of institutions concerned with the furtherance of the activities of professional 'scientists'. These social and intellectual transformations occurred during Maxwell's lifetime, and – as with his appointment to the newly established professorship of experimental physics at Cambridge in 1871 – helped shape his career. In his inaugural lecture at Cambridge in October 1871, Maxwell made allusion to the theologically neutral, progressive and secular optimism of contemporary scientific endeavour being generally seen as expressing the 'material spirit of the age' (*SP*, **2**: 251).

Maxwell certainly believed in scientific progress, the 'approach . . . to the truth' as he described it in his Cambridge inaugural lecture (*SP*, **2**: 251–2); nor was his research on electromagnetism divorced from laboratory physics and the practical demands of telegraphy, as in his participation in work on electrical standards in London in the 1860s and later at the Cavendish Laboratory. But there is another dimension to his science. He adhered to the traditional aspirations of the natural philosopher, apparent in his concern with the philosophical foundations of physics, which had played an important role in physical theorising since the work of Descartes, Newton and Leibniz in the seventeenth century.[11] In expounding the rationale of his physical theories he regularly and deliberately raised questions of analogy and intelligibility, discussed the meaning of the concepts of matter and force, and examined the conceptual status of the laws of mechanics and thermodynamics. Though his sense of duty prompted him to accept university appointments and to shoulder responsibilities within the scientific community, his ultimate commitment was that of a natural philosopher. As Lewis Campbell, his lifelong friend and biographer expressed it, 'with sacred devotion [he] continued in mature life the labours which had been his spontaneous delight in boyhood'.[12]

Maxwell held a traditional view of the cultural values of science, and lacked sympathy with the secularism of contemporary rhetoric urging the progressive advance of science and the authority of the scientific professional.[13] He placed emphasis on moral and theological values, rather than the advancing edge of scientific expertise, the rebuttal of orthodoxy, and the benefits of material progress. He associated the study of science with the growth of individual understanding, which he saw as leading to an awareness of the boundaries of scientific knowledge. In his inaugural address at King's College

London in 1860 he declared that 'we, while following out the discoveries of the teachers of science, must experience in some degree the same desire to know and the same joy in arriving at knowledge which encouraged and animated them'. The study of science encouraged a theological obligation: 'We shall find that it is the peculiar function of physical science to lead us . . . to the confines of the incomprehensible' (*LP*, **1**: 670, 674).

Maxwell urged caution in accepting claims that science could alone provide intellectual enlightenment. He pointed to the unexamined assumptions that lay behind the optimism voiced in the rhetoric of some of his contemporaries, which looked to a clerisy of scientists to rebut obscurantism and establish a scientific and secular enlightenment. 'The peculiar faith required of a positivist', he told Mark Pattison in April 1868, 'is in the universal validity of laws, the form of which he does not yet know' (*LP*, **2**: 360). But as he explained to Cambridge colleagues in February 1873, he questioned the assumption that 'the physical science of the future is a mere magnified image of that of the past'. Nature was not completely transparent to unrestricted calculation, as could be seen by recognising that there are limits to complete knowledge and perfect predictability: 'only in certain cases', he observed, can the scientist 'predict results with even approximate correctness'. Maxwell rejects the hubris of 'a supposed acquaintance with the conditions of Divine foreknowledge' (*LP*, **2**: 820–3); there are boundaries to scientific understanding.

II Formative influences

II.1 The scientific culture of Edinburgh

In autobiographical remarks written in 1873 Maxwell recollected his early delight in the forms of regular figures and curves, and his view of mathematics as the search for harmonious and beautiful shapes.

> I always regarded mathematics as the method of obtaining the best shapes and dimensions of things; and this meant not only the most useful and economical, but chiefly the most harmonious and the most beautiful.[1]

His early interest in geometry is apparent in a letter written to his father in June 1844, when he was just 13, describing his construction of 'a tetrahedron, a dodeca hedron, and 2 more hedrons that I dont know the wright names for'.[2]

Maxwell's starting-point in mathematics was geometry, leading to his first scientific paper, on the mechanical description of ovals, an interest probably inspired by his strong aesthetic sense of the harmony of geometrical forms. His first studies in physics, which began in 1847, were of the chromatic effects exhibited by crystals and mechanically strained glass when viewed in polarised light. He later described these phenomena as exhibiting 'gorgeous entanglements of colour' (*LP*, **2**: 563), again displaying his aesthetic appreciation of nature. Writing with the benefit of personal acquaintance, his biographer William Garnett noted his continued fascination with colour harmonies.[3] Maxwell's first interests in mathematics and physical science were stimulated by his sense of harmony and beauty, the subtle and transient relations of form and colour.

These early scientific and mathematical concerns were initiated and fostered by his family circle. His father John Clerk Maxwell and his uncle John Cay had interests in the sciences; the family circle encouraged youthful scientific enthusiasm. His cousin Jemima Wedderburn (who was to become well known as a watercolourist and illustrator) recorded an episode from his childhood, where the boy Maxwell is seen helping to inflate a large hot air balloon designed by Hugh Blackburn (whom Jemima was to marry in 1849, the year he became Professor of Mathematics at Glasgow University).[4] John Clerk Maxwell participated in the scientific life of Edinburgh, through membership of the Royal Society of Edinburgh and the Royal Scottish Society of Arts, associations which were to prove important to Maxwell's initiation into the scientific community.

During the winter of 1845–6, while still at school at the Edinburgh Academy, Maxwell accompanied his father to meetings of these scientific societies. He became aware of the work of David Ramsay Hay, a prominent Edinburgh decorative artist, who was engaged in several studies that aimed to reduce beauty of form and colour to mathematical principles. As Hay explained in his *Principles of Beauty in Colouring Systematized* (1845), 'Visible beauty is of two kinds; – the first arising from harmony of Form – the second from harmony of Colour'. He considered the beauty of colours and geometrical forms to be similar to the harmony of music, and hence to be explained by mathematical laws analogous to the principles of musical harmony.[5] His *Nomenclature of Colours* (1848) exhibited an elaborate system of colour plates distinguishing variations of colour, and his terminology for the classification of colours – 'hues, tints and shades' – was later adapted by Maxwell (see Chapter III.1). Hay's current interest in constructing 'the perfect egg-oval' led Maxwell to his first mathematical paper, a method of drawing oval curves.

Hay's interest in the aesthetics of curvilinear figures arose from his work in the decorative arts; and it was for purposes of decoration that he was led to devise a 'machine for drawing the perfect egg-oval', which he described to the Royal Scottish Society of Arts on 9 March 1846. In his *First Principles of Symmetrical Beauty* (1846) he discussed the drawing of oval curves by adapting Descartes' method of drawing an ellipse, where a string of any given length is held by pins at the foci, and a pencil stretches the string to trace out the ellipse.[6] Maxwell developed this method in his paper 'Observations on circumscribed figures having a plurality of foci, and radii of various proportions' (*LP*, **1**: 35–42), written early in 1846.

Although Maxwell did not discuss the mathematics of ovals in his paper, his method (in which – in the first case discussed – a thread is wrapped m and n times respectively round pins placed at two given foci) is consonant with the mathematical properties of Cartesian ovals: a tracing pin describes an oval so that m times its distance from one focus together with n times its distance from another focus is equal to a constant quantity.[7] Maxwell drew a variety of ovals with two and three foci, and succeeded in drawing ovals of up to five foci.

John Clerk Maxwell assiduously promoted his son's youthful effort, invoking his relation with James David Forbes, Professor of Natural Philosophy at Edinburgh University. Writing on 6 March 1846 Forbes assured the anxious father that his son's paper was 'very ingenious – certainly very remarkable for his years'; and after consulting his colleague Philip Kelland, Professor of Mathematics at Edinburgh, he reported that the paper set out 'a new way of considering higher curves with reference to foci'.[8] As Forbes stated in the 'remarks' he prepared in giving an account of the paper to the Royal Society of Edinburgh on 6 April 1846, the method established 'the identity of the oval [drawn by Maxwell's technique] with the Cartesian oval' (*SP*, **1**: 2). The paper was not therefore merely an ingenious method of description of ovals – on a par with Hay's work – but was of mathematical interest; hence Forbes' willingness to present Maxwell's construction, embedded in erudite remarks of his own, before the audience of scientific Edinburgh.

Forbes' allusion to Descartes prompted Maxwell to study with critical attention the ovals in Descartes' *Geometrie.* Writing to his father in April 1847 he claimed that 'I have identified Descartes' ovals with mine'. With youthful confidence he asserted that he could 'disprove' Descartes' discussion of optical reflection, the oval being supposed as the concave surface of a mirror (*LP*, **1**: 62). Some twenty years later, writing to the mathematician James Joseph Sylvester, he recollected this early interest in ovals, observing that in discussing the optical properties of ovals Descartes had made 'erroneous statements about reflexion' (*LP*, **2**: 296).

Maxwell was still at school at this time, and had there formed a friendship with Peter Guthrie Tait, who was to become his closest scientific correspondent. Tait later recalled that they discussed 'numerous curious problems' and exchanged manuscripts.[9] Among the papers referred to by Tait are sets of propositions on ovals, probably written early in 1847, which demonstrate the growth of Maxwell's mathematical knowledge and the imprint of his reading since his scientific debut the previous year. Drawing freely on Euclid's *Elements* he gives a systematic discussion of the geometrical and optical properties of ovals (*LP*, **1**: 47–61). His discussion of the geometry of ovals and description of the Cartesian conjugate ovals, which he terms 'meloid' and 'apioid' (perhaps from the Greek for apple and pear, suggested by the shape of the curves), strongly indicates his close reading of Descartes.

At around this time, April 1847, Maxwell's interest had been aroused in another pattern of inquiry. His uncle John Cay took Maxwell and his close school friend (and later biographer) Lewis Campbell on a visit to the laboratory of the Edinburgh experimental optician William Nicol. Maxwell later recalled that

> I was taken to see [William Nicol], and so, with the help of 'Brewster's Optics' and a glazier's diamond, I worked at polarization of light, cutting crystals, tempering glass, etc.[10]

He began to experiment on the chromatic effects of polarised light in doubly refracting materials, crystals and mechanically strained glass, investigations which began in earnest in autumn 1848 (*LP*, **1**: 97–103). Entranced by these 'gorgeous entanglements of colour', he looked to the writings of the leading authority on these phenomena, Sir David Brewster. The Scottish optical community – Hay, Nicol and Brewster – provided the stimulus and context for this interest.

Maxwell entered the University of Edinburgh in autumn 1847, attending Sir William Hamilton's class in Logic, Forbes' in Natural Philosophy, and Kelland's in Mathematics. Forbes divided his class into three divisions, and while Tait (who entered the University at the same time as Maxwell, but remained for only one session before going up to Cambridge) joined the first division, where knowledge of the calculus was required, Maxwell was content with the middle division of the class. In the 1847–8 session Forbes paid special attention to mechanics, with some discussion of pneumatics, heat and the steam engine; and with a substantial number of lectures on astronomy.[11] Maxwell's work in mathematics prospered: an essay written for Kelland in March 1848 shows a grasp of analytical geometry,[12] a flexing of mathematical muscles displayed to greater effect in the work of that summer, a comprehensive memoir on the theory of rolling curves, curves generated like the cycloid by one curve rolling on another.

The draft of this paper (*LP*, **1**: 74–95) differs most markedly in its organisation from the paper as published the following year (*SP*, **1**: 4–29). In the draft he classifies his examples of rolling curves into orders, genera and species, using categories of classification that had been discussed by Hamilton in his Logic lectures delivered during the previous session. Maxwell had taken detailed notes on these lectures,[13] which he found 'far the most solid' (*LP*, **1**: 69) of the courses he attended. He explained his classification of rolling curves to Lewis Campbell in September 1848:

> One curve rolls on another, and with a particular point traces out a third curve on the plane of the first, then the problem is: Order I. Given any two of these curves, to find the third.

> Order II. Given the equation of one and the identity of the other two, find their equation.
> Order III. Given all three curves the same, find them. In this last order I have proved that the equi-angular spiral possesses the property, and that no other curve does. This is the most reproductive curve of any.
> (*LP*, **1**: 96)

The strategy of this paper, written when Maxwell was only 17 years of age, points to the style of the work of his mature years. The paper is comprehensive in its grasp of the subject and literature, making reference to classic work as well as to recent publications; and it is systematic in scope, demonstrating that 'all the more notable curves may be . . . linked together in a great variety of ways, so that there are scarcely two curves, however dissimilar, between which we cannot form a chain of connected curves' (*SP*, **1**: 6).

During his second session, in 1848–9, he continued to attend lectures in mathematics and took Hamilton's class on Metaphysics, which fostered his enduring interest in philosophy (see Chapter II.3). He now entered the first division of Forbes' class. During this session Forbes again concentrated on mechanics, with some attention to the physical properties of bodies and the physics of heat, and with substantial coverage of optics. First-division students studied texts such as Poisson's *Mechanics* and George Biddell Airy's account, in his *Mathematical Tracts*, of the undulatory theory of light,[14] the major innovation in physics in the 1820s and 1830s. Forbes was responsive to current teaching at Cambridge (see Chapter II.2), and paid some attention to topics with a mathematical cast such as the wave theory of light.

Maxwell's own private reading and research continued unabated. At the end of his second session at Edinburgh he resumed his work on the chromatic effects of polarised light, studying the phenomena of induced double refraction in mechanically strained glass (*LP*, **1**: 117–19). Very ambitiously, he began to develop an explanation for these photoelastic effects by appealing to the theory of elastic solids (*LP*, **1**: 125–7). In looking to work on elasticity he was probably guided by Forbes, who had recently presented a paper on the measurement of the extensibility of solids to the Royal Society of Edinburgh.[15] Maxwell himself began to experiment on elasticity, measuring the coefficients of elasticity of rods and wires, and to consider the determination of the compressibility of water (*LP*, **1**: 128–31, 177–83), a problem that had engaged the attention of Oersted and Regnault and indeed of Forbes, who in February 1848 discussed his work on this topic with William Thomson, Professor of Natural Philosophy at Glasgow.[16]

Maxwell began to draft a systematic paper embracing the elasticity of solids and the chromatic effects of strained glass in polarised light (*LP*, **1**: 133–83). The extraordinary thrust of his paper 'On the equilibrium of elastic solids', presented to the Royal Society of Edinburgh in February 1850 (*SP*, **1**: 30–72), derives from his thorough grasp of complex material and his presentation of a mathematical theory of elastic solids; this provides the theoretical framework for his discussion of elasticity and photoelasticity. His conceptual grasp and control of the material demonstrates the depth of his mathematical reading and his confidence in tackling a subject at the forefront of research. The paper is remarkable – astonishing for an 18-year-old – and unique in its breadth of coverage. It joins discussion of the general mathematical theory of elastic bodies to its application to special cases of elastic deformation (taking examples which had been discussed in the literature), and concludes with an account of photoelasticity. Maxwell displays the comprehensive command of the material and systematic presentation characteristic of his mature work.

The mathematical theory of elasticity had been the subject of considerable recent effort by the most notable French mathematicians, Navier, Poisson and Cauchy, and Maxwell shows some grasp of their work. But he rejected their grounding of the theory of elasticity on hypotheses of interactions between the molecules of the elastic bodies, appealing instead to the approach adopted in a recent memoir by George Gabriel Stokes, 'On the theories of the internal friction of fluids in motion, and of the equilibrium and motion of elastic solids', presented to the Cambridge Philosophical Society in 1845. Stokes of course took it for granted that the phenomena of elasticity and the motion of fluids were produced by their 'molecules' and 'particles' (he uses both terms), but his theory is presented as a 'geometrical' model independent of physical hypotheses about 'molecular forces'; he assumed the independence of the shape and volume of a system of points moving in 'elements' of space.[17]

Following Stokes, Maxwell rejects any theory of an elastic solid based upon physical assumptions about the 'laws of molecular action' of the substances (*SP*, **1**: 71). He therefore discounts the central force theories of Navier and Poisson, which sought – in Poisson's words – to explain elasticity by a 'physical mechanics'[18] grounded on molecules acting at a distance, in favour of phenomenological 'axioms' (the 'results of experiments') which state relations between the pressure and compression of elastic solids. The equations deduced from these axioms, he claims, 'agree with all the laws of elasticity which have been deduced from experiments' (*SP*, **1**: 31). Maxwell's approach here resembles the method he adopted in his first memoir on field theory, 'On Faraday's lines of force' (see Chapter IV), where the distinction between a 'geometrical' model and a 'physical' hypothesis is basic to the argument of the paper. This distinction was characteristic of contemporary Cambridge mathematics, being drawn by Airy in his *Mathematical Tracts* and in papers by Thomson, as well as by Stokes.[19] Even before his Cambridge education, Maxwell had become familiar with and had adopted seminal features of the Cambridge mathematical style. His paper 'On the equilibrium of elastic solids' owes its origins to the concerns of Edinburgh experimental physics (the work of Brewster and Forbes), but its presentation to a joining of Edinburgh experimental inquiry and the Cambridge mathematical method.

Another important element of his physics owed its genesis to Forbes' interest in colour theory. Forbes introduced Maxwell to experiments on the mixing of colours in 1849, at a time when he was reviewing the problem of the classification of colours.[20] Writing to Forbes about his own work in May 1855, Maxwell recalled that 'I remember seeing several of your experiments in the summer of 1849 . . . your experiments . . . were to me the origin of the whole enquiry' (*LP*, **1**: 301–2). These experiments consisted in observing the hues generated by adjustable coloured sectors fitted to a rapidly spinning disc, using tinted papers supplied by D. R. Hay. Maxwell's work on colour vision was shaped by the attempts by Hay and Forbes to provide a method and nomenclature for the classification of colours, and especially by Forbes' method of experimentation. His interest broadened to include the problem of colour-blindness, with which he had become familiar through the investigations of another member of the Edinburgh scientific community, George Wilson (see Chapter III.1). While theoretical problems form the dominant theme in Maxwell's career, these concerns interacted productively with his commitment to the practice and values of experimental science, an interest which had its origins in his background in Edinburgh science.

II.2 Cambridge: the Mathematical Tripos

Maxwell arrived in Cambridge as an accomplished mathematician and physicist, whose intellectual development had been shaped by Edinburgh traditions of experimental science, the mechanical arts and philosophy. According to his friend Tait, then at Peterhouse, 'he brought to Cambridge, in the autumn of 1850, a mass of knowledge which was really immense for so young a man, but in a state of disorder appalling to his methodical private tutor . . . William Hopkins',[21] whose pupil he became in October 1851.

The opinion of his Edinburgh mentor mixed plaudits with some criticism. Forbes wrote to William Whewell, Master of Trinity College, in September 1850, recommending Maxwell to Whewell's attention with the caution that 'he is not a little uncouth in manners, but withal one of the most original young men I have ever met with, & with an extraordinary aptitude for physical enquiries'. After Maxwell's migration from Peterhouse to Trinity in December 1850 Forbes wrote again to introduce him to Whewell, observing that 'he is a singular lad, & shy', but 'very clever and persevering'; subsequently reminding Whewell that Maxwell was 'the author of some papers of great promise in the Edinburgh Transactions'. When Maxwell had gained a scholarship at Trinity in April 1852 after his success in the Previous Examination, Forbes told Whewell that

> I am . . . aware of his exceeding uncouthness, as well Mathematical as in other respects . . . I thought the Society and Drill of Cambridge the only chance of taming him & much advised his going . . . . But he is most tenacious of physical reasonings of a mathematical class, & perceives there far more clearly than he can express. This (in my experience) is a rather rare characteristic, & I should think he might be a discoverer in those branches of knowledge, as indeed he already partly is.[22]

Forbes and Hopkins (if Tait's recollection was accurate) were in agreement that Maxwell's exuberant genius required the discipline of systematic and ordered mathematical education. This is what study for the Mathematical Tripos could provide, and Hopkins' teaching provided the essential preparation for the examination.

Cambridge mathematics education had gone through various changes in the first half of the nineteenth century. To the mathematical reformers of early nineteenth-century Cambridge the fluxional and geometrical mathematics deriving from Newton's *Principia* appeared to be a hindrance to mathematical progress. The continental tradition of analysis, of Johann Bernoulli and Lagrange, which developed the Leibnizian differential calculus, seemed to offer a more satisfactory system of methods and notation.

In Section I of the *Principia*'s first book Newton justifies his infinitesimal arguments by the method of prime and ultimate ratios of nascent and vanishing finite quantities approaching their limits. He declares that he did not suppose indivisible magnitudes but finite limit ratios: 'by the last ratio of vanishing quantities you must understand not the ratio of quantities before they vanish, nor that afterwards, but that with which they vanish'. Moreover, he claims that the determination of this limit is a 'geometrical problem'. He makes it clear that he did not suppose 'ultimate' magnitudes to be 'indivisible', and hence in Euclidean terms incommensurable with finite magnitudes, but to be quantities 'diminishable without limit'.[23] These arguments, and especially Newton's claim that his demonstrations were justified by geometry, figured prominently in subsequent discussions of the foundations of the calculus.

The most important and systematic justification of Newton's calculus is Colin MacLaurin's *Treatise of Fluxions* (1742). MacLaurin wrote in response to the criticisms of Newton's fluxional method advanced by George Berkeley in *The Analyst* (1734), and also in an effort to contest the Leibnizian calculus. He urged that infinitesimal magnitudes (as employed in the Leibnizian calculus) should be rejected because their use is not 'founded on accurate principles agreeable to ancient geometry'; by contrast, he claimed that Newton's method of fluxions 'requires the supposition of no quantities but such as are finite and easily conceived', a method that is based on 'the ratio which is the limit of the various proportions of finite increments'.[24]

The geometrical interpretation of the foundations of the calculus became dominant in Cambridge and in the Scottish universities, along with an emphasis on the study of geometry itself. The study and exegesis of the first six books of Euclid's *Elements of Geometry* was central to the mathematical curriculum: when Maxwell sat the Cambridge Previous Examination in March 1852 the prescribed curriculum included Euclid's *Elements* Books I and II.[25] The doctrine of abstraction, that knowledge is gained through a comparison of experiences, played an important role in Scottish discussions of the foundations of knowledge. Robert Simson's famous note elucidating the distinctions between points, lines and surfaces in terms of the comparison of experiences of external objects, was reproduced in John Playfair's 1795 edition of Euclid's *Elements*, an edition that was familiar to Maxwell.[26] From Colin MacLaurin and David Hume to John Leslie and Dugald Stewart, claims that the geometric concept of points, lines and surfaces could be justified by 'contemplating an external object . . . by successive acts of abstraction', and that the 'geometry of the Greeks . . . [provides the] finest specimens of logical deduction', were important in establishing the status of geometry within a liberal education. The view that powers of reasoning were fostered 'by the study of Greek geometry . . . not by the practice of the modern methods', sustained claims for the conceptual clarity and methodological purity of the Newtonian calculus.[27]

These arguments on the mathematical and metaphysical strictness of geometry were dispelled for Cambridge reformers by Lagrange's theory of analytical functions. In his *Théorie des Fonctions Analytiques* (1797) Lagrange took the derivative (rather than the differential) as the fundamental concept of analysis, defining the derivatives of a function as the coefficients of the terms in its expansion as a Taylor series. The first, though abortive, attempt to bring the 'foreign notation' and the Lagrangian concept of the derivative to Cambridge mathematics had been made by Robert Woodhouse in his *Principles of Analytical Calculation* (1803). Woodhouse's work had little impact, but the foundation in 1812 of the Analytical Society by Charles Babbage, John Herschel and George Peacock had greater effect in bringing about mathematical reform at Cambridge. They argued that geometrical and physical illustrations should be replaced by abstract principles of analysis; this meant the Lagrangian analytical theory of derivatives – 'the more correct and natural method of derivatives' – in place of Newtonian fluxions and ultimate ratios. Peacock maintained that the limit was a mysterious concept, though he conceded that it could be made rigorous; while the Lagrangian formulation could provide clear foundations for the calculus, which would then 'be established upon principles which are entirely independent of infinitesimals or limits'. He declared that 'Taylor's Theorem . . . exhibits the whole theory of the Differential Calculus'.[28]

By the 1820s the Lagrangian theory of the derivative was established in the curriculum, coexisting with but not displacing the Newtonian doctrines of ultimate ratios and limits. Indeed, in the early 1850s, the first part of the Tripos examination, devoted to the more elementary parts of mathematics, included 'the 1st, 2nd and 3rd Sections of *Newton's Principia*; the Propositions to be proved in Newton's manner'.[29] Candidates would therefore be familiar with Newton's method of prime and ultimate ratios, his theory of centripetal forces, and his treatment of the motion of bodies in conic sections. In giving a proof of one of the 'problems' that in 1853 he set for solution in the *Cambridge and Dublin Mathematical Journal* (the house journal of Cambridge mathematics, edited by William Thomson), Maxwell makes appeal to a geometrical argument invoked by Newton in a proposition on centripetal force (*SP*, **1**: 79; *LP*, **1**: 234).

To supplement his *Introduction to Dynamics* (1832), in 1846 Whewell published an edition of the first three sections of *Principia*, which gave an exposition of the text adhering to Newton's synthetic proofs. Consonant with his espousal of geometrical mathematics he urged, in his *Doctrine of Limits* (1838), that the concept of a limit was unavoidable in the exposition of the differential calculus; and he gave a wholly unrigorous, intuitive definition of limits: 'what is true *up to* the limit is true *at* the limit'.[30]

Cauchy's 'arithmetical' approach to the calculus, based on a rigorous definition of limits by finite differences made as small as desired, did not find favour in Cambridge, where the Lagrangian approach prevailed. William Hopkins' notes on the 'Differential & Integral Calculus', copied by Maxwell (and before him by Stokes), reflect this approach. A discussion of Taylor's theorem leads to the definition of a derivative as the coefficients of the terms in the Taylor series; and he appended a further definition of a derivative (in a manner characteristic of Cambridge texts) by appeal to a limiting ratio, yet without a rigorous definition of a limit.

> [W]e may give a definition of the diff. co. different from that previously given, and depending on the idea of a limiting ratio. Let x be the independent variable of which u is the function so that $u = f(x)$. Let x receive an increment h so that u becomes $f(x+h)$ [$= u'$]. The ratio of the increment is $(u' - u)/h$. The value of the numerator depends upon, and vanishes with, that of the denominator, and that limiting value to which the ratio approximates, as the increment h is continually diminished is the diff. coeff^{t} of u.[31]

Lagrange's influence did not foster an interest in study of the foundations of the calculus.

The major and distinctive feature of the more advanced part of the Mathematical Tripos was emphasis on the study of 'mixed mathematics', including mechanics, hydrodynamics, astronomy and the theory of gravitation, and geometrical and physical optics.[32] Whewell's *Treatise on Dynamics* (1823), J. H. Pratt's *Mathematical Principles of Mechanical Philosophy* (1836, $_2$1845) and Airy's *Mathematical Tracts* (1826, $_2$1831, $_3$1842) were important Tripos texts. In the 1830s Whewell had successfully argued for the introduction of more physical subjects – the wave theory of light (treated in the second and third editions of Airy's text), heat, electricity and magnetism. This enlarged curriculum did not ultimately satisfy his view of mathematics as providing a liberal education; and in the 1840s the Tripos was again reformed, its analytical focus being tempered by a more traditional concern with geometry, and its scope restricted to the more established branches of mathematical physics. In

1849 the Board of Mathematical Studies recommended that 'the Mathematical Theories of Electricity, Magnetism, and Heat, be not admitted as subjects of examination',[33] in line with current practice, though questions on these topics continued to be set in the papers for the Smith's Prizes for which the high 'wranglers' in the Tripos examination competed.

The topics covered in Maxwell's Cambridge notes reflect Hopkins' concern with 'mixed mathematics': notes on mechanics, hydrostatics and hydrodynamics, astronomy and geometrical optics[34] give evidence of his study of these topics. While Tait later recollected that '[Maxwell] to a great extent took his own way',[35] formal study with Hopkins was demanding. In 1854, the year Maxwell sat the Tripos examination – graduating second wrangler to E. J. Routh, but subsequently being ranked joint Smith's Prizeman with Routh – 'mixed' physical questions outnumbered 'pure' analytical questions by three to two in the Tripos; and the Smith's Prize papers set by James Challis (Plumian Professor of Astronomy), Whewell, and Stokes (Lucasian Professor of Mathematics) were overwhelmingly geometrical and physical in emphasis.[36]

The scope of 'mixed mathematics' as taught in Cambridge in the 1850s owed much to the tradition and style of mechanics deriving from *Principia*, whose study formed an important element in physics and mathematics education in the university. The *Principia* is celebrated for Newton's statement of his three laws of motion as the basis of a systematic science of mechanics. Newton's statement of the laws of motion, and his elaboration of mechanical principles, did not however stand as immutable principles of physics. The second law of motion that 'change in motion is proportional to the motive force impressed and takes place following the straight line along which that force is impressed',[37] has an especially tangled history. Newton's statement of the law, as a relation between an impulsive force and a change of momentum, contrasts with the familiar modern version of the law, which states a relation between a continuous force and a rate of change of momentum. Newton conceives the action of a continuous force as the cumulative effect of an infinite number of discrete infinitesimal force-impulses impressed in successive instants of time. He supposes that forces can be 'impressed once and instantaneously, or successively and gradually'; a continuous force is conceived as the limit of a sequence of force-impulses, and Newton's statement of the second law applies to both versions of the law.[38]

In 1750, however, Leonhard Euler stated a 'new principle of mechanics' which he claimed to be the 'general and fundamental principle of all mechanics'; this new principle is the equation $f = m\, dv/dt$. The equation itself was generally familiar in the early eighteenth century, and Euler considered his principle to be new because of its generality: the principle of linear momentum applies to all kinds of mechanical system, whether discrete or continuous.[39] However, in Euler's writings, as in the work of D'Alembert and later Lagrange, Newton's laws are not stated. In Lagrange's *Mécanique Analytique* (1788) D'Alembert's principle and the principle of virtual work are given pride of place.

In contrast to the presentation offered by French authors, notably by Poisson in his *Traité de Mécanique* (1811), Whewell questioned the shift from the primacy accorded to Newton's three laws of motion. In his *Elementary Treatise on Mechanics* (1819) and his *Treatise on Dynamics* (1823) the science of dynamics is considered to be 'the science of Force, considered as producing or altering motion'; force is defined as 'that which causes change in the motion of a body'.[40] Whewell aimed to give the concept of force and the laws of motion special prominence.

William Hopkins' notes on dynamics, which were copied by Stokes and Thomson in the 1840s and by Maxwell in the 1850s, reflect Whewell's stress on the primacy of the Newtonian laws of motion. Whewell's statement of Newton's second law of motion in his *Treatise on Dynamics* is echoed by Hopkins:

> *Second Law of Motion.* If any number of forces act on a material point in motion, the intensity and direction of each force will be the same as if the point were at rest.

The 'mathematical analysis of the problem' consists in calculating the effects of '[impulsive] forces during an element of the time (δt) and of combining these effects with the previously existing momentum'.[41] This follows Newton's interpretation of the second law of motion, as a relation between impulsive force and a change in momentum.

This Newtonian version of the law was presented in the texts of three of Hopkins' pupils: by Thomson and Tait in their *Treatise on Natural Philosophy* (1867) and by Maxwell in his *Matter and Motion* (1876). In Maxwell's words:

> Law II.– The change of momentum of the system during any interval of time is measured by the sum of the impulses of the external forces during that interval.[42]

In his exposition of analytical dynamics as the foundation of electromagnetic theory in his *Treatise on Electricity and Magnetism* (1873) Maxwell stresses the physical basis of the language of dynamics. In his view the equations of dynamics should be based on the concept of momentum, so that Newton's second law of motion, rather than Lagrangian symbols, would determine the meaning of the dynamical theory of the electromagnetic field (see Chapter VI.1).

While work with Hopkins provided the basis for study for the Tripos examination, Maxwell did attend Stokes' lectures on 'Hydrostatics, pneumatics, and optics' in the Easter Term 1853.[43] Beginning with an account of the properties of fluids and the equations of hydrodynamics, Stokes discussed waves in fluids and the theory of sound, followed by lectures on the wave theory of light. His opening lecture usually summarised the content of the course; his outline for this lecture as given in 1852 reads:

> Sketch, Foundations of hydrostatics – capillary attraction – gases and vapours – instruments – touch on hydrodynamics – Optics. Not a great while to geometrical optics; physical optics; large part of course, devote considerable time. New subject.[44]

Stokes' lectures were not considered to be an essential preparation for the Tripos examination, and from Maxwell's casual remark that he would 'look in upon Stokes' [lectures] dealing with light' (*LP*, **1**: 219) it may be judged that his commitment to attendance was not strong. Hopkins did however lecture on these topics, and indeed considered that

> It is only when the student approaches the great theories, [such] as Physical Astronomy and Physical Optics, that he can fully appreciate the real importance and value of pure mathematical science . . . .[45]

Maxwell's notebook containing Hopkins' notes on 'Hydrostatics, Hydrodynamics, & Optics' includes notes on geometrical, not physical optics. Nevertheless it may be confidently surmised that he became familiar with the broad outline of Fresnel's wave theory of light as presented in Airy's *Mathematical Tracts* (see Chapter VIII.1). But study of the mathematical theories of heat, electricity and magnetism – topics which from the 1840s were the main fields of research in mathematical physics – did not form part of the Tripos examination in the 1850s.

Maxwell's interests, however, did not remain confined by the formal curriculum. A manuscript on 'Rankine on Elasticity' (*LP*, **1**: 206–7), where he expresses Rankine's theorems in geometrical form, shows his continued interest in the theory of elasticity and a preference for diagrams over analysis.[46] He continued his interest in colour mixing, performing experiments on the mixture of coloured beams of light in August 1852 (*SP*, **1**: 144). Having been introduced to current studies of magnetism by William Thomson (*LP*, **1**:

205) he followed Thomson's work on the subject. A manuscript note on the 'Mathematical Theory of Polar Forces' (*LP*, **1**: 210–11) strongly suggests a reading of Thomson's 1851 paper 'A mathematical theory of magnetism', and his awareness of Thomson's development of Faraday's concept of the 'magnetic field' (see Chapter IV.1).

Maxwell's interest in geometry found expression in a substantial memoir in which he developed work by Gauss on the curvature of surfaces (*LP*, **1**: 239–41; *SP*, **1**: 80–114). He introduced a concept of 'lines of bending' to represent the transformations in which a surface changes its form, characteristically preferring geometrical to analytical conceptualisation. This paper was presented to the Cambridge Philosophical Society in March 1854, shortly after the Tripos examination. A month earlier he had written to Thomson, having 'entered the unholy state of bachelorhood', to declare his 'strong tendency to return to Physical Subjects . . . to attack Electricity' (*LP*, **1**: 237). This endeavour was to lead to his first paper on field theory, 'On Faraday's lines of force' (1856), where he develops a geometrical representation of the electromagnetic field in terms of 'lines of force' (see Chapter IV). There is a hint of a link between his depiction of the bending of surfaces by 'lines of bending' and his espousal of 'lines of force' in this paper. Study for the Mathematical Tripos had sharpened the tools necessary for this ambitious project.

II.3 Philosophical education: Edinburgh and Cambridge

Maxwell had a lifelong interest in metaphysics and ethics, and in the methodological problems of physical theorising. His work displays an abiding concern to establish the conceptual rationale of his physics by an appeal to philosophical argument. His reflections on the scope of physical theory, in an 1856 essay on 'Analogies in Nature' written for the Apostles club at Cambridge, indicate the character of his thought:

> The dimmed outlines of phenomenal things all merge into another unless we put on the focussing glass of theory and screw it up sometimes to one pitch of definition, and sometimes to another, so as to see down into different depths through the great millstone of the world. (*LP*, **1**: 377)

This passage, displaying a sophisticated understanding of scientific knowledge as theory-dependent, expressing the view that phenomena are intelligible only if viewed through the 'telescope of theory in proper adjustment' (*LP*, **1**: 381), and recognising that theories can be complementary in generating scientific understanding, is startling in its philosophical acuity. While Maxwell's reflections on philosophical issues bear the stamp of his originality of mind, his outlook was shaped by his reading and formal education.

At Edinburgh he attended Sir William Hamilton's classes in Logic and Metaphysics, gaining formal exposure to philosophy. But in his lectures to the Natural Philosophy class Forbes placed some emphasis on methodological issues – stressing the value of analogical reasoning and Baconian experimentalism, and urging caution in the adoption of hypotheses. In giving attention to such questions Forbes was following the tradition of his predecessors in the Edinburgh chair of Natural Philosophy.[47]

At Cambridge the curriculum of the Mathematical Tripos embodied cultural assumptions. Whewell's texts and his historical and philosophical works stressed the value of mathematical reasoning, and its cultural significance in providing evidence of certitude, order and design in nature. John Herschel's *Preliminary Discourse on the Study of Natural Philosophy* (1830) placed emphasis on the close relation between mathematics and dynamics, the science of force and motion: hence dynamics was 'at the head of all the sciences' and was capable of providing 'a certainty no way inferior to mathematical demonstration'.[48] Mathematics, quantification, and the special status of mechanics – and hence the role of 'mixed mathematics' within the curriculum – thus had cultural resonances.

Maxwell's education exposed him to the philosophical norms current in the intellectual milieu of the 1850s. As a product of the Edinburgh educational tradition, he arrived at Cambridge alerted to philosophical issues, but his writings show an engagement with the foundations of scientific knowledge unique among contemporaries (such as Thomson and Tait) whose educational background was not dissimilar to his own. His emphasis on the 'practical relation of metaphysics to physics', as he put it in 1868 (*LP*, **2**: 361), was exceptional, in his commitment of interest and breadth of culture, and in the originality of his arguments.

Maxwell's informed engagement with philosophical issues was an essential part of his scientific discourse, and will be a recurring theme in this book (see especially Chapter IX). But these discussions must be understood in relation to the scientific problems in which they are embedded, rather than as elaborations of a developing philosophical programme. Maxwell wrote as a natural philosopher. He did not participate in the vociferous and bruising public controversies which Victorian scientists conducted in the literary periodicals.

Alluding to Whewell's confrontations with David Brewster, he remarked, in a letter of July 1858 to his Trinity friend Cecil James Monro, that 'it is well I am not a Literary Man' (*LP*, **1**: 596). He published no essays that were specifically philosophical in content, and did not formally adhere to any school of opinion in philosophy. The tracing of philosophical influences is notoriously slippery, so discussion here will be focused on his education and documented from his writings, to gain a preliminary perspective on his outlook.

Hamilton's course on Metaphysics, which Maxwell attended in his second session at Edinburgh (in 1848–9), had a significant impact on his intellectual development. His lifelong friend and biographer, Lewis Campbell, attested that Hamilton's ideas 'were his habitual vantage-point'[49] in his excursions into speculative thought. Maxwell's allusions to Hamilton later in life were not wholly complimentary (*LP*, **2**: 361, 597), but there are good grounds for accepting the judiciousness of Campbell's recollection. Hamilton's lectures were based upon an extraordinary erudition; their published text – an authentic record of the lectures Maxwell attended – is dense with citations from ancient, patristic and continental philosophy. Focusing on the concerns voiced in the lectures of his academic predecessors, Hamilton's own position embraced the traditional Scottish philosophical outlook conjoined to a distinctive Kantian perspective.

Hamilton gave detailed consideration to the views of Thomas Reid, Dugald Stewart and Thomas Brown – the Scottish 'common sense' school of philosophy. These Scottish philosophers aimed to combat the corrosive and uncompromising scepticism advanced in the writings of David Hume. Hume's scepticism about the authenticity of knowledge of the external world, and his critique of the principle of causality, were seen by Reid to threaten the common sense principles which are taken for granted in the common concerns of life, 'the truth of which every man may see by attending to his own thoughts'.[50] Reid and his successors thought that such scepticism would ultimately lead (as with Hume) to atheism, and would be destructive of moral virtue.

In response to Hume's emphasis on rigorous philosophical foundations, which exposed the fragile basis of all claims to knowledge, the common sense philosophers were concerned to clarify the basis in which reasoning was commonly deployed, to demarcate its intelligibility rather than to establish its logical foundations and rigour. The status of scientific knowledge – in particular the nature of scientific reasoning, the relationship between mathematics and physics, and the status of geometry – had an important place in this philosophical enterprise. Several of these themes receive special attention in Maxwell's writings: the concern with intelligibility, the role of geometry and analogy, and the relation between mathematical and physical representations of nature.

In an essay 'On the properties of matter' (*LP*, **1**: 110–13) written for the Metaphysics class during the 1848–9 session, and which Hamilton himself preserved, Maxwell makes a number of statements which indicate the imprint of Hamilton's lectures. He begins, as was conventional in philosophical discussions of the nature of matter, with an account of the primary or essential properties of matter; these were terms current in the philosophical literature since the time of Boyle, Newton and Locke. He terms the position, spatial extension and figure of bodies 'the geometric properties of matter' because 'these three properties belong both to matter and to imaginary geometrical figures'. Maxwell's designation recalls Dugald Stewart's term, as reported by Hamilton: the '*mathematical affections of matter*'. Stewart had gone on to argue that while our first notions of extension and figure were suggested by sensory perceptions, once acquired 'the mind is immediately led to consider them as attributes of space no less than of body; and . . . becomes impressed with an irresistible conviction that their existence is necessary and eternal'.[51]

Hamilton gave this argument a specifically Kantian gloss, introducing a distinction between 'space' and spatial extension. He declares 'space' to be 'one of our necessary notions, – in fact, a fundamental condition of thought itself'; adding that the 'analysis of Kant . . . has placed this truth beyond the possibility of doubt'. To maintain the relation between cognition and material reality, which he describes as 'the great problem of philosophy', he distinguishes between 'space, as a necessary notion . . . [which] is native to the mind', and the spatial extension of material bodies; spatial extension is 'an element of existence' from which we gain 'our empirical knowledge of space'.[52] Maxwell may well have had this argument in mind when he describes geometric figures as 'forms of thought and not of matter' (*LP*, **1**: 111).

The doctrine of abstraction, that knowledge is gained through a comparison of experiences, played a seminal role in Scottish metaphysics. As Hamilton expressed the idea: 'comparison is supposed in every, the simplest, act of knowledge'.[53] Maxwell employs the doctrine of abstraction, which could have become familiar to him from philosophical and mathematical texts (the writings of Dugald Stewart and John Leslie) as well as from Hamilton's lectures, in maintaining that by the comparison of two dissimilar spheres 'we learn, by a kind of intuitive geometry, the position of external objects in three dimensions' (*LP*, **1**: 112).

Along with the doctrine of abstraction, the concepts of analogy and relation were important elements in Scottish writings on metaphysics and geometry. It is these latter concepts that are invoked in Maxwell's essay 'Are there real Analogies in Nature' (*LP*, **1**: 376–83), written for the Apostles in February 1856. There are likely echoes of the writings of Scottish luminaries such as Colin MacLaurin and Dugald Stewart in Maxwell's essay. MacLaurin had emphasised the status of mathematical knowledge as grounded upon 'the relations of things rather than their inward essences', claiming that 'our ideas of relation are often clearer and more distinct than those of the things to which they belong'. This argument helped shape the doctrine advanced by Scottish philosophers, that knowledge is analogical and relational. As Stewart expressed it, analogies implied a '*correspondence* (or, as it is frequently called, a *resemblance*) of *relations*'; the perception of such relations 'implies the exercise of reason'.[54]

In his essay on 'Analogies in Nature' Maxwell provides a general discussion of analogical reasoning in science. At the time of writing he was engaged in applying the physical analogy of the flow of an incompressible fluid as a geometrical representation of Faraday's theory of magnetic lines of force (see Chapter IV.2).

> Whenever [men] . . . see a relation between two things they know well, and think they see there must be a similar relation between things less known, they reason from one to the other. This supposes that although pairs of things may differ widely from each other, the *relation* in the one pair may be the same as that in the other. Now, as in a scientific point of view the *relation* is the most important thing to know, a knowledge of the one thing leads us a long way towards a knowledge of the other.
>
> (*LP*, **1**: 381–2)

In the final paragraph of his essay he takes up the issue described by Hamilton as 'the great problem of philosophy': 'to analyse the contents of our acts of knowledge, or cognition – to distinguish what elements are contributed by the knowing subject, what elements by the object known'.[55] To illustrate the close relation between cognition and reality, between thoughts and things, Maxwell invokes the geometric structure of crystals, 'sparkling in the rigidity of mathematical necessity'. In the regular forms of crystals geometrical relations are embodied in the physical disposition of the molecules of matter; their geometric symmetry illustrates his claim that 'the only laws of matter are those which our minds must fabricate, and the only laws of mind are fabricated for it by matter' (*LP*, **1**: 383).

The relationship between cognition and reality was highlighted in William Whewell's *Philosophy of the Inductive Sciences* (1840, $_2$1847), there described as the 'fundamental antithesis of philosophy'.

> In all cases, Knowledge implies a combination of Thoughts and Things. Without this combination, it would not be Knowledge. Without Thoughts, there could be no connexion; without Things, there could be no reality. Thoughts and Things are so intimately combined in our Knowledge, that we do not look upon them as distinct.[56]

Maxwell was certainly familiar with Whewell's work by 1855, if not earlier (*LP*, **1**: 315). In 1854 and 1855 he was a candidate for a fellowship at Trinity College, where the examination was in classics, mathematics and metaphysics. By the high standards required Maxwell's expertise in classics was deemed deficient, and he was unsuccessful at his first attempt.[57] Whewell's writings were considered essential reading for the metaphysics paper[58] – he was after all Master of the College. So it is not surprising that allusions to Whewell's ideas begin to appear in Maxwell's letters and papers in 1855, the year of his election as a Fellow of Trinity.

Whewell's essential philosophical thesis, Kantian in inspiration, is that science forms a comprehensive system of laws which, while being based on the results of experimental labour, are also both universal and necessary. His 'fundamental antithesis of philosophy' expresses the disjunction between 'Ideas and Senses, Thoughts and Things, Theory and Fact'.[59] This issue, which Hamilton termed 'the great problem of philosophy', is basic to the argument of Maxwell's essay on 'Analogies in Nature'. Whewell's and Hamilton's discussions were consilient, and very likely shaped his construal of the problems of philosophical inquiry.

The most distinctive and original feature of Whewell's *Philosophy* is his concept of '*Fundamental Ideas*'; this structures the argument of the book. These are guiding principles that regulate scientific theories, and are prior to and not directly established by experience. 'These ideas entirely shape and circumscribe our knowledge; they regulate the active operations of our minds, without which our passive sensations do not become knowledge'. The 'fundamental ideas' establish the conceptual foundations of the sciences: geometry rests on the idea of space (which is necessarily three-dimensional); mechanics depends on ideas of force, matter and cause; the idea of symmetry is basic to crystallography. Whewell does not contend that 'fundamental ideas' are innate to the intellect or are logically necessary; he considers them to be necessary in the sense that their negation could not be clearly conceived: 'necessary truths are those of which we cannot distinctly conceive the contrary'.[60] Like the Scottish philosophers, Whewell is concerned with the intelligibility of concepts, not with the questions of logical foundations and rigour.

There are many references in Maxwell's writings to 'fundamental ideas', which he also designates as 'appropriate ideas', making explicit reference to Whewell (*LP*, **1**: 315, 378, 421, 519). These 'fundamental ideas', he later explained, in a review of Thomson and Tait's *Elements of Natural Philosophy* (1873), were 'modes of thought by which the process of our minds is brought into the most complete harmony with the process of nature' (*SP*, **2**: 325). Maxwell's discussion of the harmony of thoughts and things is clearly Whewellian in inspiration; and his definition of the necessity of the 'fundamental ideas' also follows Whewell. Thus in his inaugural lecture at King's College London in October 1860 he explains that necessary truths were those 'which the mind must acknowledge as true as soon as its attention has been directed to them' (*LP*, **1**: 668).

In the essay 'Analogies in Nature' he discusses the question whether the concept of three-dimensional space is a 'fundamental idea' or merely a projection 'of our mental machinery on the surface of external things'. To resolve the issue, he considers 'whether there is anything in Nature corresponding to' the concept. He concludes that the assertion of the 'impossibility of conceiving a fourth [spatial] dimension' agrees with the 'objective truth that points may differ in position by the independent variation of three variables'. Hence he finds that in this case there is 'a *real* analogy between the constitution of the intellect and that of the external world'. The concept of three-dimensional space thus bridges the chasm between 'thoughts and things' (Maxwell uses Whewell's expression), and hence has the status of a 'fundamental idea' (*LP*, **1**: 378).

Whewell appeals to the science of dynamics to support his claim that fundamental principles, not directly established by experience, regulate scientific theorising. His explication of the *a priori* form of the laws of motion is shaped by Kant's procedure in his *Metaphysical Foundations of Natural Science.* Whewell expresses the idea of cause in terms of three axioms of causality, which when applied to causes of motion yield three *a priori* laws of forces; when reformulated in terms of the motion of bodies, these laws in turn generate the Newtonian laws of motion.[61]

Whewell's aim was to establish that the laws of motion have an *a priori* form as well an empirical content, a thesis that has no counterpart in Maxwell's discussion of the status of the laws of motion. But Whewell also made the more limited claim, that the laws of motion are not directly established by experiment. This argument is presented by William Hopkins in his Cambridge notes on dynamics, transcribed by Maxwell and before him by Stokes and Thomson.

> The Laws of motion . . . are frequently considered as having been established by experiments. It will be more correct however to consider them rather as suggested than established by any direct experiments made for the express purpose of verifying them. They must be considered as established by that inductive power of reasoning by which in fact the truth of known laws of Nature is confirmed. In this case the process is this: the fundamental laws are assumed and the motion of a material point or System of points is calculated with this assumption. The results of such Calculations are then compared with the actually observed motion of the system. It is in the perfect accordance of these results of calculation and observation that the ultimate proof of all the elementary principles assumed consists.[62]

Maxwell follows this line of argument. In a review of Whewell's writings in 1876, he explained that in Whewell's view 'the fundamental doctrines of mechanics' were supported by experiment, but 'once fairly set before the mind [are] apprehended by it as strictly true' (*SP*, **2**: 530). He echoes Whewell in claiming that establishment of the laws of motion rested on the fact that their negation could not be intelligibly conceived. He expressed the point clearly in his inaugural lecture at Marischal College, Aberdeen in November 1856:

> I maintain that as soon as we clearly understand what motion is, and that force is that which alters motion we are able to prove all the laws relating to force. (*LP*, **1**: 426)

The laws of motion were not necessary *a priori*, but were necessary in the Whewellian sense, that their contrary could not be distinctly conceived. In his text *Matter and Motion* (1876) he argued that the denial of Newton's first law of motion would be 'in contradiction to the only system of consistent doctrine about space and time which the human mind has been able to form' (see Chapter IX.1).[63]

Maxwell placed boundaries around metaphysics. 'I have no reason to believe that the human intellect is able to weave a system of physics out of its own resources without experimental labour', he declared in his inaugural lecture at Marischal College (*LP*, **1**: 425–6). Yet he insisted on the close relation between physics and metaphysics. Whewell had argued that physical discoverers differed from barren speculators 'not by having *no* metaphysics in their heads, but by having *good* metaphysics while their adversaries had bad'.[64] Maxwell echoed this view (see Chapter IX.1), and in similar vein he told his Aberdeen audience that

> I would have you remember that the men to whom we owe the greatest discoveries in mathematics and physics were metaphysicians. *They* thought it a very important thing to determine the *evidence* on which they built any law.

The two subjects were closely related: 'the greatest and most original metaphysicians have been *nourished* as it were on physical truth', and he cites Descartes and Leibniz as examples. Thus metaphysics draws on physics:

> In fact the things discussed in metaphysics are so intimately connected with the foundations of Natural Philosophy, that we have only to read a few pages of a metaphysical work to ascertain the precise limits of the author's knowledge of physical science. (*LP*, **1**: 424–5)

In this sense metaphysics was dependent on physical science. As he remarked to his Cambridge friend Richard Buckley Litchfield in March 1858, 'the chief *philosophical* value of physics is that it gives the mind something distinct to lay hold of' (*LP*, **1**: 588).

The emphasis on the intelligibility of concepts, and on clarity of expression being achieved through geometrical models, are characteristic features of Maxwell's philosophy. In his paper 'On a dynamical top' (1857), on the rotation of a rigid dynamical system, he discusses Louis Poinsot's theory of rotating bodies. He points out that Poinsot's method was based on the 'liberal introduction of "appropriate ideas", chiefly of a geometrical character'. While he clearly intended to be understood as making allusion to Whewell, he here attaches a specifically geometrical, rather than abstractly philosophical, meaning to 'appropriate ideas'. He explains that in Poinsot's method 'ideas take the place of symbols, and intelligible propositions supersede equations'. Thus ideas of a 'geometrical character' provide the basis of 'true knowledge' (*SP*, **1**: 248–50). The geometrical representation of concepts is associated with the intelligibility of physical theory, a major theme in the writings of Scottish mathematicians and philosophers. He makes the same point in his paper 'On

Faraday's lines of force' (1856), urging his 'geometrical model' of fluid flow as giving 'embodied form' to mathematical 'symbols', and hence providing a 'clear physical conception' of the theory of lines of force (*SP*, **1**: 156, 158, 187; see Chapter IV.2).

In his review of Thomson and Tait's *Elements of Natural Philosophy* he emphasises the importance of expressing physical ideas 'in appropriate words without the aid of symbols'; such 'fundamental ideas' would be 'clothed with the imagery . . . of the phenomena of the science itself'; and by way of illustration he again alludes to Poinsot's geometrical representation of dynamical concepts (*SP*, **2**: 325, 328). Clarity and intelligibility, expressed through geometrical modelling, would provide a representation of physical reality. Maxwell terms this modelling a 'mental representation of the facts' (*SP*, **2**: 360); the 'faculty of Representation', Hamilton had remarked, is 'the power the mind has of holding up vividly before itself the thoughts which . . . it has recalled into consciousness'.[65]

These remarks suggest the imprint of the philosophical ideas deployed by Hamilton and Whewell. Their expressions echo in the terms he uses in expounding his dynamical theory of physics (see Chapters VI.1 and IX.1) in his *Treatise on Electricity and Magnetism* (1873).

> The fundamental dynamical idea of matter, as capable by its motion of becoming the recipient of momentum and of energy, is so interwoven with our forms of thought that, whenever we catch a glimpse of it in any part of nature, we feel that a path is before us leading, sooner or later, to the complete understanding of the subject.
>
> (*Treatise*, **2**: 181 (§550))

III Edinburgh physics and Cambridge mathematics

III.1 Casting light on colours

In May 1855, while engaged in correcting proofs of his paper 'Experiments on colour, as perceived by the eye', Maxwell received a letter from J. D. Forbes setting out his recollection of his experiments on colour mixing, witnessed by Maxwell some six years earlier. In his reply Maxwell recalled that

> I remember seeing several of your experiments in the summer of 1849 both with whirling discs & with the colours of the spectrum. I saw white produced by three colours of the spectrum, the rest being stopped and it was from what I saw then combined with some observations of Newtons about such white light differing from sun light, that turned my attention to this subject.

Maxwell's approach to colour vision was shaped by the attempts by Forbes and D. R. Hay to provide a method and nomenclature for the classification of colours; and especially by Forbes' method of experimentation: 'your experiments . . . were to me the origin of the whole enquiry', he assured Forbes (*LP*, **1**: 301–2).

Maxwell's work on colour vision has its origins in his time as Forbes' pupil at Edinburgh University. More broadly, the work may be seen as a product of the Scottish tradition of experimental optics, and specifically an outgrowth of the interest in colour theory among Scottish scientists during the first half of the nineteenth century. This was not a subject cultivated by Cambridge mathematicians; only in the 1850s, with the work of Helmholtz, Grassmann and Maxwell himself, did the subject acquire any clear quantitative expression. The wave theory of light had been incorporated into Cambridge 'mixed mathematics' in the 1830s; but the relation of this work to colour theory seemed obscure. As Maxwell told Forbes in November 1857, it was his aim 'to get the scientific world to see the difference between a theory of colour and a theory of light' (*LP*, **1**: 571). In Scotland, by contrast, Brewster, Forbes and Hay were all concerned with problems in the theory of colours.

Maxwell's statement about the impact of reading Newton's remarks on colour mixing is of especial interest. In the *Opticks* Newton explains colour

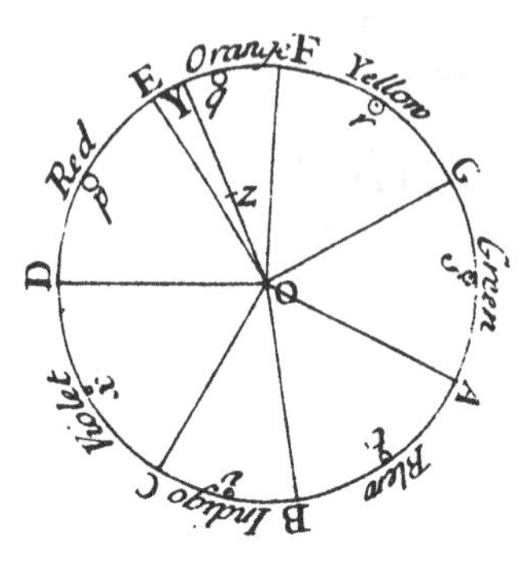

Fig. III.1. Newton's colour-mixing circle. From Isaac Newton, *Opticks* (London, $_2$1718), Bk. I, Pt. II, Pl. III.

mixing by his colour circle (Fig. III.1). The circumference of a circle is divided into arcs proportional to the lengths of the seven spectral colours. The circumference represents pure, unsaturated hues (without any admixture of white). The mixture of the spectral colours as in the sun's white light is represented by the centre of the circle *O*. At the centre of each arc he places a small circle, whose size (or weight) is proportional to the number of rays of that particular colour in any given mixture. The point *Z* represents the common centre of gravity of all the small circles, and indicates the colour compounded from a mixture of spectral colours.

Newton does however distinguish between white produced by a mixture of, say, three spectral colours and the sun's white light, which is composed of the seven spectral colours mixed in due proportion. This is the point highlighted by Maxwell in his letter to Forbes of May 1855. In his discussion of the colour circle, Newton recognises that if two of the principal colours that are opposite to one another in the circle are mixed in an equal proportion, then 'the point *Z* shall fall upon the centre *O*'. Nevertheless, he maintains that the colour thus compounded will 'not be perfectly white, but some faint anonymous Colour'. While he declares that he had failed to produce white from two colours (as Huygens had suggested might be possible by mixing blue and yellow lights), he allows the possibility of achieving it with 'a mixture of three taken at equal distances in the circumference'. Any colour thus compounded is however to be distinguished from sunlight, where 'there uses to be a mixture of all sorts of rays, and by consequence a composition of all Colours'.[1]

In his Royal Society paper 'On the theory of compound colours, and the relations of the colours of the spectrum' (1860), concerned with the mixture of spectral colours, Maxwell distinguishes between 'the optical constitution, as revealed by the prism' and the '*chromatical* properties' of a mixture of colours. He echoes Newton, who uses differences in refrangibility, determined by passing the rays through a prism, to physically differentiate between colour mixtures which may be sensibly identical. He expounds Newton's distinction between the sun's white light and white compounded from a given mixture of colours:

> Newton is always careful, however, not to call any mixture white, unless it agrees with common white light in its optical as well as its chromatical properties, and is a mixture of *all* the homogeneal colours.
>
> (*SP*, **1**: 411–12)

Here he makes reference to Newton's letters to Oldenburg in 1673, written in reply to Huygens, where Newton distinguishes between a white produced by mixing several spectral colours and the 'White of the sun's immediate light'.[2]

In explaining the mixing of spectral colours, Newton implicitly assumes the identity of the mixing rule for spectral lights and pigments. The painters' triad of primary colourants, red, yellow and blue, was introduced in the seventeenth century and soon assimilated into the scientific tradition. Newton defines the seven principal spectral colours by their refrangibility, as those colours whose rays refracted equally; but he supposes that the mixing of, say, blue and yellow spectral colours is to be understood as analogous to the mixing of blue and yellow pigments. Although he classifies spectral green as a homogeneal or simple colour, he also claims that green can be compounded from mixing the adjacent spectral colours blue and yellow: 'the yellow and blue on either hand, if they are equal in quantity they draw the intermediate green equally towards themselves in Composition'.[3] While simple spectral green is immutable and cannot be altered by further refraction, compound green, Newton claims, can be resolved by refraction into blue and yellow.

On discussing the mixing of spectral blue and yellow in his letter to Forbes in May 1855, Maxwell remarks that 'most optical writers speak confidently of the same mixture as a fine green', adding that 'in your own paper on Classification of colours you seem to be of that opinion' (*LP*, **1**: 302). In his paper 'Hints towards a classification of colours', published in the *Philosophical Magazine* in March 1849, Forbes had indeed followed Newton: 'by combining the primary colours, such as yellow and blue together, a green, not distinguishable from that of the spectrum, *except by its refrangibility*, will be formed'.

Forbes' explication of the mixture of spectral colours is based on the assumption that 'the Newtonian spectrum consists of the three primary colours, red, yellow and blue, and the three secondary, orange, green and purple'. He describes Tobias Mayer's triangular scheme for the classification of colours, proposed in 1775, based on red, yellow and blue as the primary constituents of colour. In Forbes' colour triangle, red, yellow and blue are placed at the vertices of a triangle, with the secondary colours orange, green and purple being located at the mid-points of its sides. In this triangular scheme, 'a point in the triangle may always be found which shall represent *any possible* proportional mixture of the three [primary] colours'. While continuing to maintain, with Newton, the identity of the mixing rule for lights and pigments, he does remark on the 'essential difference' between these two processes, 'of compounding rays of the spectrum and compounding pigments'. Mixing lights, he declares, is an additive process, but 'by combining pigments we do not add together *lights*'.[4]

Forbes did not however develop this insight, which was not without precedent in the literature,[5] to question the assumption of the identity of the mixing rule for lights and pigments. This conventional mixing rule was however soon challenged by Helmholtz, in a paper 'On the theory of compound colours' published in 1852. In experimenting on the mixing of spectral colours he found 'that yellow and blue do not furnish green'. This 'contradicts in the most decided manner the experience of all painters during the last thousand years' in mixing blue and yellow pigments. He argued that while the mixture of coloured lights is an additive process, pigment mixing is subtractive; pigments act as filters to light reflected from interior layers below the surface. In attempting to produce white from mixing just two spectral colours, he found that only 'yellow and indigo-blue', the colours whose combination had been generally assumed would produce green, could be combined together to give 'pure white'.[6]

Forbes had already gone some way towards Helmholtz's discovery. In January 1849 he had experimented with a rapidly spinning disc fitted with adjustable coloured sectors (rather than, as was conventional, painting the colours on the disc), the tinted papers being supplied by D. R. Hay. On observing the hues generated, Forbes found (as he recorded in his laboratory notebook) that 'Yellow and blue only, equal, produce a yellow grey or citrine – *never green*'. Referring to these experiments in his letter to Maxwell in May 1855, he claimed that 'before Helmholtz, or I believe any one else', he had found 'that the mixture of yellow and blue, under these circumstances at least [the blending of coloured papers on a whirling disc], does not produce green; you yourself being a witness to what I then tried'.[7]

In response Maxwell remarked that

> I must consider you as the first person who ever attempted to make green by spinning blue and yellow, in spite of all experiments of the kind formerly described, for if any one had really tried it he would have been struck by the result, and would have made *your* discovery.
>
> (*LP*, **1**: 302)

He added a supplementary note to the proofs of his paper 'Experiments on colour, as perceived by the eye, with remarks on colour-blindness', published in the *Transactions of the Royal Society of Edinburgh* in 1855. He acknowledged the 'experiments of Professor J. D. Forbes, which I witnessed in 1849', experiments which established 'that blue and yellow do not make green, but a pinkish tint, when neither prevails in the combination', and he affirmed that this 'result of mixing blue and yellow was, I believe, not previously known' (*SP*, **1**: 145–6). But while Helmholtz had questioned the identity of the mixing rule for pigments and spectral colours, Forbes' experiments with coloured papers on a whirling disc did not lead him to this conclusion. Indeed, as Maxwell reminded him, in his paper of March 1849 on the classification of colours he had affirmed the view of 'most optical writers' that blue and yellow hues when combined generated 'a fine green' (*LP*, **1**: 302).

Witnessing Forbes' experiments had aroused Maxwell's interest in colour theory; and in August 1852, during the Cambridge summer vacation and independently of Helmholtz's work, he mixed coloured beams of light. His subsequent experiments confirmed that, in mixing spectral hues, 'blue and yellow do *not* make green' (*SP*, **1**: 144, 244). But his major sequence of experiments, reported in his paper 'Experiments on colour, as perceived by the eye', involved an improvement of Forbes' use of Hay's adjustable tinted papers fitted to a rotating disc. Maxwell devised a colour top, a 'chromatic teetotum', which he had constructed by an Edinburgh instrument maker, in which he added a second set of adjustable coloured papers of smaller diameter to the first set, so that colour comparisons could be made between two different sets of colours (*LP*, **1**: 284–5).

> The method of experimenting consisted in placing before the eye of the observer two tints, produced by the rapid rotation of a system of discs of coloured paper, arranged so that the proportions of each of the component colours could be changed at pleasure. (*LP*, **1**: 287)

The circumference of the top was divided into 100 graduations, so that the proportions of each colour, in a given combination, could be read off. The sectors of coloured paper were adjusted so that the tints of the outer and inner circles were 'perfectly indistinguishable, when the top has a sufficient velocity of rotation' (*SP*, **1**: 128). He constructed colour equations expressing the equality between the sectors in the outer and inner circles. To achieve objectivity he enlisted the aid of observers, finding that there was little variation in the observations made by different individuals. This quest for objectivity was an essential feature of his method. He told Thomson in March 1854 that 'by using the aid of others in every experiment . . . very exact results may be obtained about the equivalence of two colours as regards the eye'. Such agreement between observers was 'the condition of a science of sensible colour independent of individual peculiarities' (*LP*, **1**: 246). The use of colour equations gave the argument a clear quantitative expression.

Maxwell's interest in the theory of colours broadened beyond colour mixing to include the problems of colour vision and colour-blindness. This interest was well established by March 1854, when he informed Thomson that 'Colour *as perceived by us* is a function of three independent variables at least three are I think sufficient' (*LP*, **1**: 245).

Maxwell probably had in mind the three-receptor theory of colour vision expounded by Thomas Young in his *Course of Lectures on Natural Philosophy* (1807), a text which had been reissued, edited by Philip Kelland, in 1845. As Maxwell described the theory, Young maintained that there are 'three distinct modes of sensation in the retina, each of which he supposed to be produced in different degrees by different rays' (*SP*, **1**: 136). Writing in January 1855 to George Wilson, an Edinburgh acquaintance who was currently investigating the problems of colour-blindness, Maxwell explained that according to this theory

> it is not necessary to specify any given colours as typical of these sensations. Young has called them red, green, and violet; but any other three colours might have been chosen, provided that white resulted from their combination in proper proportions. (*LP*, **1**: 269)

Young's theory of colour vision incorporated his scheme of red, green and violet as the primary spectral colours; and he described a colour triangle showing that all other colours, including white, could be generated from combinations of these primaries. In breaking the analogy between the painters' triad of primary colourants and spectral hues, this choice of primaries had been judged to be 'a singular opinion' by Forbes in 1849.[8] By January 1855 Maxwell had adopted Young's scheme, but emphasised that the crucial point in the selection of primaries was their combination, in due proportion, to form white.

Having adopted Young's three-receptor theory of colour vision, Maxwell accepted the suggestion, made by Young and later by John Herschel, that colour-blindness was caused by the absence of one of the three receptors.[9] He explained the theory to Wilson:

> Suppose the absent structure to be that which is brought most into play when red light falls on our eyes, then to the Colour-blind red light will be visible only so far as it affects the other two sensations, say of blue and green. (*LP*, **1**: 273)

Colour-blindness was the result of an 'absent sensation' in the retina (*LP*, **1**: 289), and he suggested that 'it is the red *sensation* which is wanting, that is, that supposed system of nerves which is affected in various degrees by all light, but chiefly by red' (*SP*, **1**: 138). He tested this hypothesis by recording observations with the colour top made by colour-blind observers.

Maxwell was also able to give a new and comprehensive account of the classification of colours. He adopted the theory recently proposed by Hermann Grassmann, that there are three variables of colour vision (spectral colour, intensity of illumination, and the degree of saturation). Grassmann demonstrated that this method of representing colour could be expressed by the position and magnitude of loaded points on Newton's colour circle.[10] Maxwell terms these variables 'hue' (representing the wavelength), 'intensity' or 'shade' (representing brightness), and 'tint' (representing the gradations of purity) (*LP*, **1**: 268–9; *SP*, **1**: 131), adapting terms from Forbes, Grassmann and Hay.[11]

He shows that these colour variables can be graphically represented on a colour diagram (Fig. III.2) which incorporates Young's triangular scheme, Newton's colour circle, and Grassmann's classification of colours (*LP*, **1**: 270; *SP*, **1**: 154). The three colour sensations red, green and violet are placed at the vertices of a triangle, so that a point within the triangle 'will be the position of the given colour, and the numerical measure of its intensity will be the sum of the three primitive sensations' (*LP*, **1**: 270). A point *W* within the triangle denotes white. The variables 'hue' and 'tint' are represented on the diagram by angular position with respect to *W* and distance from *W*, and the 'intensity' is represented by a coefficient. Grassmann's use of three colour variables and Young's triangular scheme of colour primaries 'are capable of exact numerical comparison'; and he declares that 'the relation between the two methods of reducing the elements of colour to three becomes a matter of geometry' (*SP*, **1**: 131, 135). This was an original synthesis of ideas in the theory of colours.

The experiments on colour vision, reported in his 1855 paper 'Experiments

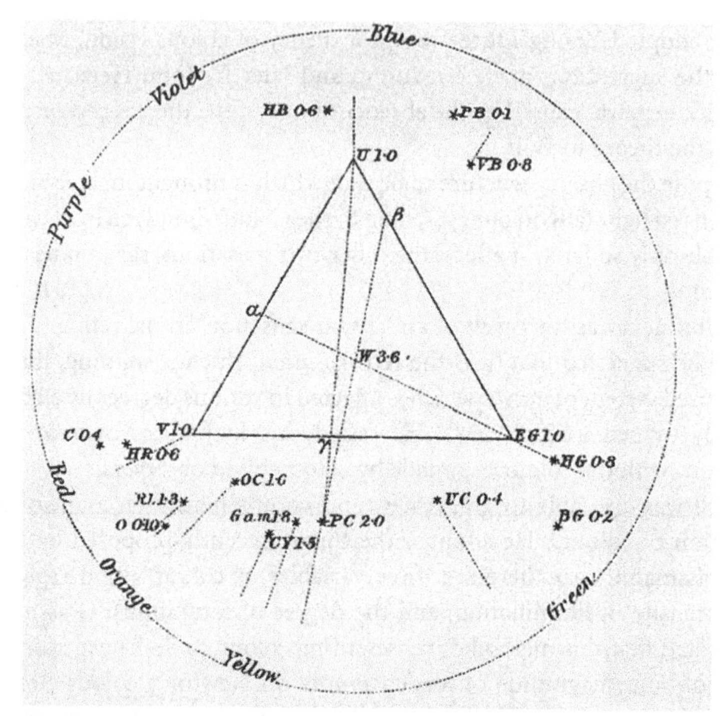

Fig. III.2. The colour diagram, described in 'Experiments on colour, as perceived by the eye' (1855), represents the relations of the coloured papers used in the experiments with the colour top. Three coloured papers, vermilion (*V*), emerald green (*EG*), and ultramarine (*U*), are taken as the standard colours, and placed at the verticles of the equilateral triangle. The other coloured papers – carmine (*C*), red lead (*RL*), orange orpiment (*OO*), orange chrome (*OC*), chrome yellow (*CY*), gamboge (*Gam*), pale chrome (*PC*), mixed green (*UC*), Brunswick green (*BG*), verditer blue (*VB*), and Prussian blue (*PB*) – are arranged in their relation to the standard colours. The point *W* within the triangle denotes the position of white. Lines drawn through *W* to the different colours establish their sequence, which corresponds to the order of the spectral colours (though the spectrum is deficient in the purples lying between ultramarine and vermilion). The distance of the colours from *W* represents the purity (degree of saturation) of the coloured papers. The numbers are the coefficients of intensity (brightness) of the colours. The letters *HR*, *HB*, and *HG* denote Hay's red, blue, and green papers. From *SP*, **1**: Pl. I facing p. 154.

on colour, as perceived by the eye', were made with the colour top, where light is reflected from the coloured papers; greater immediacy required the direct mixing and comparison of spectral colours. Writing to Forbes in November 1857, he remarked that

> Coloured papers and spinning tops though capable of far greater accuracy than most spectrum experiments convey no absolute facts about definite colours. (*LP*, **1**: 571–2)

This was because, as he explained in his 1855 paper, the 'colours on the discs do not represent primary colours at all, but are simply specimens of different kinds of paint' (*SP*, **1**: 127). Hence the 'equations are . . . between the colours of certain pigments' (*LP*, **1**: 568). For this reason, beginning in 1852, he designed a series of 'colour boxes' to mix and compare spectral colours so as to make direct measurements. The difficulty was to achieve accuracy in the experimentation. As he told Forbes:

> I have had some experience of spectrum mixing, as I have made 3 different sets of apparatus which all succeeded partially. I began in 1852 with one 3 feet long and a water prism. In 1855 I made one 7 feet long with a glass prism and arrangements for seeing two mixtures of pure colours spread uniformly over two contiguous fields. In 1856 I made a reflecting portable apparatus for showing the phenomena roughly to strangers.

He planned to develop a new apparatus to exhibit 'mixtures in juxtaposition', aiming at an improvement of Helmholtz's methods (*LP*, **1**: 569).

This he began the following year (*LP*, **1**: 600–2); and the instrument as finally perfected enabled mixtures of spectral colours to be directly compared with white light (Fig. III.3 and *SP*, **1**: 420–2). This work was reported in his Royal Society paper 'On the theory of compound colours' (1860), his last major contribution to the subject. But he continued to experiment and to develop his 'colour box' (*LP*, **1**: 709), notably in the improved manufacture of the slits by which the light was admitted to the instrument (*LP*, **2**: 608).

In his paper 'On the theory of compound colours' he sets out a comprehensive statement of his theory of colour vision, based on accurate quantitative data on spectral colours obtained from a calibrated colour box. Firmly grounded on Grassmann's vectorial approach, in which (as Maxwell put it) 'the rule for the composition of colours is identical with that for the composition of forces in mechanics' (*SP*, **1**: 418–19), and the exposition of Newton's colour circle, the paper vindicates Young's theory of colour mixing and colour vision. Indeed, he was able to plot distribution curves of the luminosity of each standard colour as a function of wavelength, representing the physiological response mechanism in the eye. Plotting curves for himself and his wife (enlisted in the experiments after their marriage in June 1858), he was able to show how each primary colour stimulates a sensation across the spectrum,

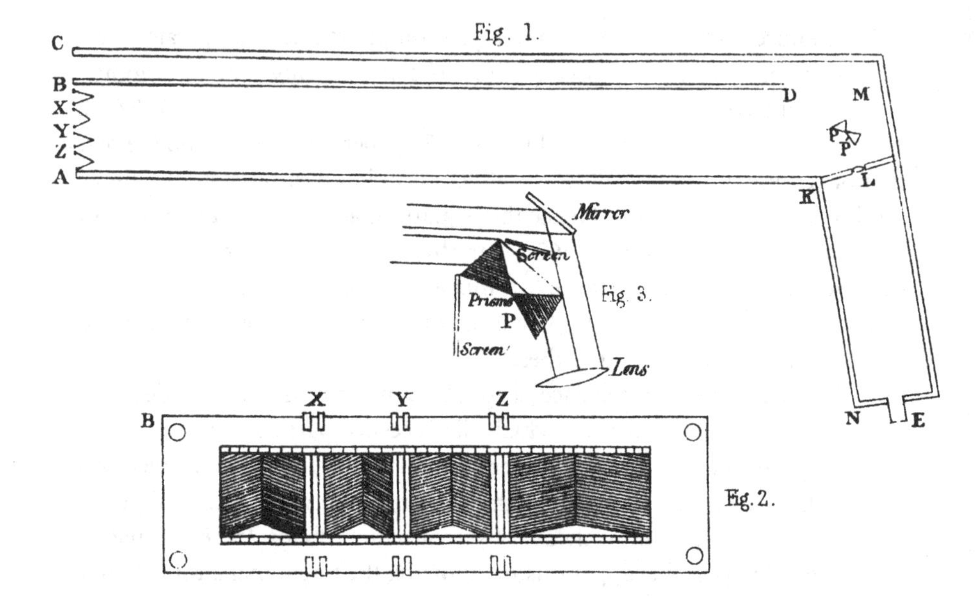

Fig. III.3. Maxwell's 'colour box' for making mixtures of spectral colours, described in 'On the theory of compound colours' (1860). The instrument (Fig. 1) consisted of two closed wooden boxes, *AK* being five foot (1.5 metres) long, *KN* two feet (60 centimetres) long, *BD* being a partition in the long box. *E* is a vertical slit, *L* a lens. Figure 3 shows the two equilateral prisms *P* adjusted so that when light is admitted at *E* a pure spectrum is formed at *AB*. A mirror at *M* is adjusted to reflect light from *E* to *BC*. Figure 2 shows the brass frame and slides at *AB*, the brass knife-edges forming three slits *X*, *Y*, *Z* which were adjusted to coincide with red, green, and blue portions of the spectrum formed by light from *E*. The colour seen at *E* from light admitted at *X*, *Y*, *Z* would be a compound of red, green, and blue light, in proportions depending on the breadth of the slits, read off on a graduated scale, and the intensity of the light which enters them. This mixture of light could be compared with white light admitted through *BC* to the mirror at *M*, and reflected to *E*. The slits *X*, *Y*, *Z*, admitting red, green, and blue spectral colours, were adjusted by an operator till the observer sees the prism illuminated by white light of the same hue and intensity as the white light reflected by the mirror. From *SP*, **1**: Pl. VI facing p. 444.

providing experimental support for Young's three-receptor hypothesis (*SP*, **1**: 433–4).

During the 1850s the study of colour theory was, in Britain, very much within the Edinburgh experimental tradition of optics – that of Brewster,

Forbes, Hay and Wilson. Maxwell's early papers were published in Edinburgh journals. Yet he wished to interest Thomson and Stokes, Cambridge-educated mathematical physicists, in this work. Writing to his father after meeting Thomson in Glasgow in September 1855, he reported enthusiastically that 'I got a long letter from Thomson about colours and electricity. He is beginning to believe in my theory about all colours being capable of reference to three standard ones' (*LP*, **1**: 326). The rather dubious and marginal status of this work in the eyes of Thomson and Stokes can however be judged from their correspondence early the following year. Thomson asked Stokes

> Have you seen Clerk Maxwell's paper in the Trans. R. S. E. on colour as seen by the eye? Are you satisfied with the perfect accuracy of Newton's centre of gravity principle on w^{h} all theories & nomenclatures on the subject are founded? That is to say do you believe that the whites produced by various combinations, such as two homogeneous colours, three homogeneous colours, &c, are absolutely indistinguishable from one another & from solar white by the best eye? . . . Are you at all satisfied with Young's idea of triplicity in the perceptive organ?

In response Stokes, a leading authority on the undulatory theory of light and the discoverer of fluorescence in 1852, replied that 'I have not made any experiments on the mixture of colours, nor attended particularly to the subject'.[12]

But by November 1857 Stokes had familiarised himself with Maxwell's work, complimenting the younger physicist that his

> results afford most remarkable and important evidence of the theory of 3 primary colour perceptions, a theory which you and you alone so far as I know have established on an exact numerical basis.[13]

This, for the Cambridge mathematical physicist was the key: exact numerical data expressed in terms of colour equations that could be manipulated algebraically, rather than mere experimental illustrations of a speculative theory. Moreover, in the 1850s, theories of triplicity in optics were under a cloud.

In his 'New analysis of solar light' (1834)[14] David Brewster had rejected Newton's explanation of the refraction of sunlight and his theory of colour. Brewster maintained that sunlight consists of only three primary colours, red, yellow and blue. He claimed that rays of these colours are dispersed through the entire spectrum, their mixture, in various proportions, producing the colours of the Newtonian spectrum. This theory was based on experiments on the absorption of light; and though initially (and tentatively) endorsed by John Herschel, it came under attack in the late 1840s at the hands of Airy and Whewell. The theory was decisively refuted by Helmholtz in 1852, who showed that Brewster's experiments were vitiated by observational errors.[15] Brewster characteristically staged a rearguard defence of his theory, witnessed by Maxwell, at the Glasgow meeting of the British Association for the Advancement of Science in September 1855. But as Maxwell reported to his father, 'Stokes made a few remarks, stating the case not only clearly but courteously' (*LP*, **1**: 325).

Maxwell's introduction of exact numerical measurements and colour equations into the discussion of colour vision fully satisfied Stokes, and the Cambridge circle moved to secure concrete public approbation for his work. In June 1859 Stokes and Whewell nominated Maxwell for a Royal Medal of the Royal Society, 'for his Mathematical Theory of the Composition of Colours, verified by quantitative experiments, and for his Memoirs on Mathematical and Physical subjects'. Thus Maxwell's colour theory was applauded as mathematical and based on numerical measurements. Another nomination, also unsuccessful, was made the following year. But a nomination in May 1860 by Stokes and William Hallowes Miller, the Cambridge professor of mineralogy, for the Rumford Medal (which was awarded especially for studies of light and heat), for Maxwell's 'Researches on the Composition of Colours, and other Optical Papers', met with success.[16] By this time, encouraged by Stokes as Secretary of the Royal Society (*LP*, **1**: 619–22, 632), he had submitted his paper 'On the theory of compound colours' to the Society. Indeed, in February 1860 he was invited to read the paper as Bakerian Lecturer, though it transpired that as he was not (until May 1861) a Fellow, he was ineligible for the appointment.[17] The work on colour vision established Maxwell's reputation as an experimental physicist working in a mathematical style: Edinburgh physics informed by the values of Cambridge mathematics.

III.2 On Saturn's rings

In 1859 Maxwell published a memoir *On the Stability of the Motion of Saturn's Rings*, a revise of the essay which won him the Adams Prize of the University of Cambridge in 1857. In the introduction to the memoir he explained that

> when we contemplate the Rings from a purely scientific point of view, they become the most remarkable bodies in the heavens, except, per-

> haps, those still less *useful* bodies – the spiral nebulæ. When we have actually seen that great arch swung over the equator of the planet without any visible connexion, we cannot bring our minds to rest. We cannot simply admit that such is the case, and describe it as one of the observed facts in nature, not admitting or requiring explanation. We must either explain its motion on the principles of mechanics, or admit that, in the Saturnian realms, there can be motion regulated by laws which we are unable to explain. (*SP*, **1**: 291)

Here he powerfully evokes the intellectual challenge of a problem in dynamics, the examination of the various models which could provide workable representations of the motion and stability of the rings of Saturn. This bore directly on Laplace's classical work on celestial mechanics; but, he remarked, 'our curiosity with respect to these questions is rather stimulated than appeased by the investigations of Laplace [on the rings of Saturn]' (*SP*, **1**: 292). The subject offered Maxwell an opportunity to establish his reputation with a contribution to mathematical astronomy in the Cambridge style.

The Adams Prize had been established in 1848 by members of St John's College in honour of John Couch Adams' prediction of the existence of the planet Neptune. On the first three occasions it had been set the subjects were on celestial mechanics, but the prize had attracted few candidates and had only been awarded once. For the Adams Prize for 1857, James Challis (as Plumian Professor of Astronomy) was determined to excite greater interest among Cambridge graduates who alone were eligible to compete. Writing in February 1855 to William Thomson, Professor of Natural Philosophy at Glasgow and a former Fellow of Peterhouse, to inform him of his appointment to serve as an examiner for the prize, Challis explained the problem of finding a suitable topic.

> The subjects hitherto proposed have been astronomical, – but they may be in 'other branches of Natural Philosophy' & in 'Pure Mathematics'. We have not hitherto been successful in inducing competitors to come forward. . . . I fear that Cambridge mathematicians have no taste for investigations that require long mathematical calculations. I should be glad if you can suggest some subject that will be more likely to attract candidates.

He enclosed a list of four possible topics. Two of these were problems in celestial mechanics, and one on the aberration of light (this last being a topic of special interest to Challis). The third subject in his list, 'An investigation of the perturbations of the forms of Saturn's Rings, supposing them to be fluid', was the one to which he gave preference. He told Thomson that this problem 'acquires an interest on account of the singular conclusions Otto Struve has recently come to respecting the approach of the inner Ring to the ball of Saturn'.[18]

The problem of the rings of Saturn had excited considerable recent interest among astronomers. A dark 'obscure ring' interior to the two familiar bright rings had been first observed by the Harvard astronomer George Phillips Bond in 1850.[19] On a visit to Europe the following year he had discussed his discovery with colleagues. Visiting Pulkovo Observatory (near St Petersburg) in August 1851, he took the opportunity to observe Saturn in the company of Otto Struve. He recorded in his Journal that he perceived that Struve 'was seeing the new ring for the first time & with entire certainty – I suspected so before he spoke'. He reported a discussion the following day with Struve and his father Wilhelm Struve, where the view was advanced that the ring system 'is in process of change'.[20]

Bond's discovery of a previously unobserved ring prompted Otto Struve to undertake a comprehensive series of observations and an exhaustive literature search, reported in a substantial memoir on the dimensions of the rings of Saturn. He maintained that the newly discovered ring was in fact a 'recent formation'. Reviewing two centuries of observations of the ring system he concluded that

> the inner edge of the interior bright ring is gradually approaching the body of the planet while at the same time the total breadth of the two bright rings is constantly increasing . . . [and] that during the interval which elapsed between the observations of J. D. Cassini and those of Sir William Herschel, the breadth of the inner ring had increased in a more rapid ratio than that of the outer ring.[21]

If it were the case that the rings were changing in form over time, then this would support the hypothesis (as Challis implied in stating his suggested prize problem) that the rings were fluid rather than solid.

Bond had made the acquaintance of the Cambridge scientific luminaries Whewell, Stokes and Challis himself in July 1851, before visiting Russia; and on his return to England in October he paid Challis a visit at the Cambridge Observatory, also meeting Adams and Whewell.[22] While Bond did not record the nature of his discussions with the Cambridge astronomers, it would be surprising if the problem of Saturn's rings passed without comment. Certainly by February 1855, having read Otto Struve's memoir, Challis was keen to set an Adams Prize subject on the question.

Responding to Challis' request for suggestions for possible subjects for the prize, Thomson proposed (among other topics) an investigation of the elasticity of solids, a problem of special interest to him at the time. This Challis considered to be 'a subject somewhat removed from the general tenor of Cambridge mathematics'. He favoured an astronomical subject, and again expressed his preference for 'the question about Saturn's Rings'.[23] He proceeded to draw up a draft on 'The mechanical stability of Saturn's Rings' for Thomson's approval. Thomson suggested that the title be amended to 'The Motions of Saturn's Rings', on the grounds that 'it may perhaps be found that the Rings do not possess mechanical stability', a possibility raised by Struve's paper.[24] This became the title heading the published notice, dated 23 March 1855, advertising the competition for the Adams Prize for 1857 (*SP*, **1**: 288).

In advocating his choice of subject for the prize, Challis had explained to Thomson that

> I quite think that a definite result respecting the stability of the forms of the Rings may be arrived at supposing no other forces to be concerned with the mutual action of the parts than that of gravity. As soon as the friction of the parts, or a resisting medium enters into the account, we have a cause operating to produce permanent change. In the instructions to the candidates I have separated the part of the Problem which may admit of a definite answer, from that which can hardly be answered except upon gratuitous hypotheses. The latter part may give rise to speculation and conjecture, which it may not be useless to encourage.[25]

The discussion of such hypotheses did indeed play a role in Maxwell's attempt to answer Struve's claim that there had been a change in form of the rings over time. In considering friction as a factor disturbing the stability of the rings, he was led to questions of gas viscosity, and ultimately began to investigate the kinetic theory of gases (see Chapter V.1).

The problem of Saturn's rings was, as Challis had implied, consonant with 'the general tenor of Cambridge mathematics'. Whewell had set a Smith's Prize examination question in February 1854 (the year Maxwell was awarded the prize), requiring that candidates 'Shew that a fluid may revolve in a perfect annulus, like Saturn's ring. How does it appear that Saturn's ring is not a rigid body?'.[26] The second part of the question involved reference to the discussion of the rings of Saturn in Chapter 6 of Book III of Laplace's *Traité de Mécanique*

Céleste, where he had established that the motions of a uniform solid ring were dynamically unstable. Laplace concluded that the rings could be irregular solid bodies whose centres of gravity did not coincide with their geometrical centres.[27]

The first part of Whewell's question, on the rotation of a fluid ring, was probably prompted by Joseph Plateau's suggestion that the appearance of the rings of Saturn was analogous to the effect of rotation on a sphere of oil immersed in a mixture of alcohol and water, where the sphere 'is transformed into a perfectly regular ring'. He noted that 'the heavens exhibit to us . . . a body of a form analogous to our liquid ring . . . [this was] Saturn's ring'. The rupture of this fluid ring, Plateau remarked, served as an 'image in miniature of the formation of the planets' according to Laplace's 'nebular' cosmogony, where planets were supposed to have been formed by the condensation of gaseous matter surrounding the primeval sun.[28] Laplace himself had suggested that the satellites and rings of Saturn were formed by the condensation of gaseous matter in the planet's atmosphere.[29]

It is not clear when Maxwell began work on the Adams Prize problem. By July 1856, when he left Cambridge to take up his appointment as Professor of Natural Philosophy at Marischal College, Aberdeen, he was fully engaged on his essay. He remarked to his friend R. B. Litchfield that he had 'been giving a portion of time to Saturn's Rings which I find a stiff subject but curious, especially the case of the motion of a fluid ring' (*LP*, **1**: 411). A comment on the stability conditions, and further discussion of the fluid ring in a letter the following October (*LP*, **1**: 417), show that he was well advanced in the work by this time. The essay was completed and submitted to the examiners (according to the terms of the competition) by 16 December 1856; Maxwell's was the only submission.[30]

The argument of Maxwell's essay 'On the Stability of the Motion of Saturn's Rings' follows the terms of reference framed by the examiners. The essay is divided into two parts, the first being concerned with the motion of a rigid ring, the second with the motion of a fluid ring or a ring formed of disconnected particles. The mathematical argument rests on potential theory, Taylor's theorem, and Fourier analysis, all methods familiar to a Cambridge wrangler. He begins with the classic work of Laplace, seeking to determine the conditions under which the rotation of a solid ring would be stable. The basic mathematical technique derives from Laplace's treatment of potential theory in the *Mécanique Céleste*. He determines the potential at the planet due to the ring: 'We have . . . to determine the forces which act between the ring and the sphere, and this we shall do by means of the *potential*, *V*, due to the ring' (*LP*, **1**: 446). Obtaining the equations of motion for the rotation of the ring about its centre of gravity, he derives conditions under which a uniform motion of the ring would be possible.

He then proceeds to consider the effect of disturbances on the motion of the ring, seeking whether they would be merely periodic or sufficiently small as to maintain dynamical stability, or would 'produce a displacement which would increase indefinitely and derange the system altogether'. Expanding the potential to first order by Taylor's theorem, he obtains 'Equations of the motion when slightly disturbed'. To investigate these deviations from a steady state he approximates and linearises the differential equations of the motion of the rings, and reduces the equations of motion to a biquadratic equation

$$An^4 + Bn^2 + C = 0,$$

where n stands for the operation d/dt (*LP*, **1**: 448–50). The coefficients A, B and C establish whether the motion of the ring would be stable or unstable.

He sets out five conditions for values of n, concluding 'that to ensure stability all the roots must be . . . pure impossible [imaginary] quantities'.

> That this may be the case both values of n^2 must be negative, and the condition of this is
>
> 1st That A, B & C should be of the same sign
>
> 2nd that $B^2 > 4AC$.
>
> When these conditions are fulfilled a periodic disturbance is possible. When they are not both fulfilled the motion cannot be permanent.
>
> (*LP*, **1**: 451)

Thus, if all real roots and all real parts of the complex roots are negative, the rings would be stable. In his review of the published memoir on *Saturn's Rings* George Biddell Airy, the Astronomer Royal, commended Maxwell's mathematical argument 'as an interesting example of a beautiful method, applied with great skill to the solution of the difficult problems which follow'.[31] Maxwell was to employ the same method in establishing stability conditions in his 1868 paper 'On governors' (*SP*, **2**: 106).[32]

In the Adams Prize essay he goes on to apply these stability conditions to determine the circumstances under which the rotation of a solid ring would be stable, finding that if a solid ring were supposed lopsided, conditions of stability would be attained, though 'the necessary amount of this inequality . . . must be so enormous as to be quite inconsistent with the observed appearance of the rings' (*LP*, **1**: 442). Nevertheless, his discussion of the stability conditions was not free from error, as Challis noticed and tried (without success) to remedy. Maxwell erred in establishing the equations for the gravitational potential of the ring; and this entailed that a uniform solid ring would be stable, contrary to Laplace's demonstration. Because of a slip, Maxwell failed to draw this conclusion; the consequence of this compensating error was that he inferred that 'the motion is unstable' in the case of a uniform solid ring:

> The result of the theory of a rigid ring shows not only that a perfectly uniform ring cannot revolve permanently about the planet, but that the irregularity of a permanently revolving ring must be a very observable quantity. (*LP*, **1**: 455)

This was therefore in accordance with his earlier observation that '[Laplace] proves most distinctly (Liv III Chap VI) that a solid uniform ring cannot possibly revolve about a central body in a permanent manner' (*LP*, **1**: 442). Only when he reworked the argument the following August did he perceive the source of his error, and conclude, on valid grounds, that a uniform solid ring would be unstable (*LP*, **1**: 533).

In the second part of the essay he turns to the case of a fluid ring. Here 'every particle of the ring is to be regarded as a satellite of Saturn'. The different parts of the ring are now considered as being capable of independent motion; hence 'we must take account of the attraction of each portion of the ring as affected by the irregularities of the other parts' (*LP*, **1**: 443, 456). Assuming the ring initially to be uniform but subject to small disturbances, he considers radial and tangential forces to act on an element of it. He finds that if the ring were assumed to be at rest 'the whole ring would collapse into satellites'; but that 'when we treat the question dynamically . . . we are able to understand the possibility of the stable motion of a fluid ring'. The effect of the motion of the ring is to compensate the tangential force produced by the action between the parts of the ring; any disturbance in the ring produces four sets of waves in the plane of the ring, as well as two sets of waves oscillating normal to the plane. He therefore concludes that

> the same tangential force, which would, of itself, break up the ring by increasing its irregularities, becomes converted by the motion of the ring into a cause which actually tends to diminish the irregularities and to reduce the ring to uniformity. (*LP*, **1**: 444)

As in the first part of the essay, he linearises the equations of motion and finds the characteristic roots of a biquadratic equation for the angular velocity (relative to the velocity of the ring) with which the system of waves travels round the ring. On discussing the question (raised by Struve) of the possible change in form of the rings over time, he considers the effect of disturbing causes on the stability of the rings. These are the friction of the rings; and an external force due to the irregularites of the planet, the mutual attraction of the satellites, or the effect of the irregularites on neighbouring rings (*LP*, **1**: 468–73).

In calculating the effect of long-continued disturbances on a system of rings he takes 'advantage of two general principles of Dynamics'. The first of these is 'the principle of the Conservation of Angular Momenta' (*LP*, **1**: 472). This principle was not generally presented as a fundamental dynamical axiom at the time, certainly not in Cambridge texts (see Chapter II.2). But in his study of rotation in his paper 'On a dynamical top' (1857) Maxwell drew on the explication of the principle in a contemporary Cambridge paper,[33] describing the conservation of angular momentum as forming part of 'our stock of appropriate ideas and methods' (*SP*, **1**: 249–51), an allusion to Whewell's philosophy (see Chapter II.3).

The second general principle is that of the conservation of energy, and again his usage and terminology reflect contemporary practice. The measure of *vis viva* (mv^2) by mechanical 'work', defined as the integral of a force with respect to distance; the interconversion and quantitative equivalence of heat and work, as established by Joule; Helmholtz's formulation of the energy principle as a mechanical theorem; and the elaboration of energy physics in the 1850s by Thomson and W. J. M. Rankine, with Rankine introducing the term 'potential energy', brought about a transformation in the scope of the science of physics.[34] Maxwell expresses the energy of the ring system in terms of 'half the vis viva of the system [V]' [kinetic energy] and 'the potential energy of the system due to Saturn's attraction [P]'. Noting that if the motion of the system was an

> angular rotation, and if no loss of power took place on account of internal friction we should have $V+P=$ const. But if there be loss of power by internal friction then $V+P$ will continually diminish.

On considering the rotation of the ring system he finds that 'the more the rings spread out from one another . . . the less will be the value of $(V+P)$'. Hence 'the ultimate effect of internal friction is to make the outer rings extend farther from the planet and the inner rings come nearer to it' (*LP*, **1**: 472–3).

The dynamical argument implied that Struve's conclusion, that the ring system had changed in form over time, could be correct.

> The result of a long-continued series of disturbances among the rings [would be] . . . that the exterior rings would recede from the planet and the interior ones approach towards his surface. This perhaps is the only one of our results which has been observed, or believed to have been observed.

He concludes the essay by suggesting that observations made with the actual telescopes used by the old astronomers might resolve the question, as to whether the changes apparently observed in the rings (by Struve) were due to 'this continual source of decay' arising from the friction of the rings, or to 'the improvement of telescopes' (*LP*, **1**: 476–7).

Maxwell was awarded the Adams Prize on 30 May 1857, and his essay was returned to him bearing the comments of Challis and Thomson. He soon began to revise his work for publication. Writing to Thomson (with whom correspondence on the subject was now at last possible) he drew attention to Struve's memoir; and he assured Thomson, who had queried his discussion of the issue,[35] that 'the encroachment of the inner ring seems very certain and not due to the improvement of telescopes' (*LP*, **1**: 531).

During the summer and autumn of 1857 Maxwell revised the argument of his essay. The course of these revisions can be followed from his letters to Thomson, Challis, Tait and Campbell (*LP*, **1**: 527–41, 553–67, 573–84). The first major point to be clarified was the error in his treatment of a uniform solid ring. He soon spotted his mistake (*LP*, **1**: 533), restructuring his argument in much the form as it appears in the published memoir (*SP*, **1**: 307–10). He now established unambiguously that a uniform solid ring would be unstable; and that only in the case of a solid ring loaded with a heavy particle where the ratio of the mass of the particle to the mass of the ring is as '82 to 18' would a solid ring be stable (*LP*, **1**: 533–5).

Turning to the problem of a fluid ring, he reported to Thomson in November 1857 that he had 'abolished my off hand theory of the attractions of a thin fluid filament affected by waves'. His new calculation established that 'the liquid continuous ring is doomed' (*LP*, **1**: 553–4). He then discussed the conditions in which the parts of a broken ring could form a permanent ring of satellites orbiting the planet. He considered the case of a 'cloud of aerolites' or 'brickbats', finding the condition of stability to be that 'the mean density of the cloud must be less than 1/330 of that of Saturn'. However, Laplace had shown that for the outer and inner parts of the ring to have the same angular velocity 'Saturn must be not more than 1.3 times as dense as the ring'. Hence a ring formed of a cloud of particles 'could not revolve with uniform angular velocity (See Laplace) so we are driven to a plurality of rings with independent angular velocities' (*LP*, **1**: 554, 566). This became the conclusion of his memoir on *Saturn's Rings*: the ring system consists of independent, concentric rings of disconnected satellites.

> The final result, therefore, of the mechanical theory is, that the only system of rings which can exist is one composed of an indefinite number of unconnected particles, revolving round the planet with different velocities according to their respective distances. These particles may be arranged in series of narrow rings, or they may move through each other irregularly. In the first case the destruction of the system will be very slow, in the second case it will be more rapid, but there may be a tendency towards an arrangement in narrow rings, which may retard the process. (*SP*, **1**: 373)

To facilitate understanding of the operation of systems of waves in a ring of satellites, he devised a model 'for the edification of sensible image worshippers'. By adjusting the position of each satellite in the model, 'the ring of satellites may be thrown into waves of any length which travel round the ring' (*LP*, **1**: 576, 579). This model provides a visual representation of Maxwell's explanation of the structure and stability of the rings of Saturn. Displayed to the Royal Society of Edinburgh in April 1858 (*LP*, **1**: 593; *SP*, **1**: 286–7), it illustrates Maxwell's mathematical analysis in the concrete form of 'a very neat model of my theoretical ring, a credit to Aberdeen workmen' (*LP*, **1**: 584). The abstractions of Cambridge mathematics were rendered visual, and transformed into Scottish physical realism (see Plate I, p. *58*).

III.3 Physics and metrology

In his inaugural lecture at Marischal College, Aberdeen in November 1856 Maxwell emphasised that

> Nothing that we can say or think here can escape from the ordeal of the measuring rod and the balance. All quantities must be exact quantities . . . so that we have a most effectual means of discovering error, and an absolute security against vagueness and ambiguity.
>
> (*LP*, **1**: 425)

Plate I Model (1858) illustrating the motions of the satellites constituting the rings of Saturn.

The measurement of quantities was therefore at the heart of a physical investigation. In a draft written the following year, intended for his Aberdeen lectures, he made the point that measurement required reference to standards: in order 'to compare things measured by different persons it is necessary to assume a standard of measure'. The measurement of physical quantities generated precision, and required standards of measure; these were not merely the arbitrary usages of mankind but represent fundamental features of nature:

> those who make up their minds to study Nature with measuring rod time-piece and weights will find that these arbitrary and perhaps inaccurate standards are intended to represent something uniform and independent of any individual man, which depends on an ancient decree and is preserved by the power of Nature which neither a new decree nor new actions of Nature could restore if it were destroyed.
> (*LP*, **1**: 520)

Units of length, time and mass provide the basis for standards of measurement; while physical standards represent the fixed and immutable order of nature.

Metrology and standards expressed more than scientific values. In his famous address on 'Molecules' to the meeting of the British Association for the Advancement of Science in September 1873 (see Chapter VIII.2), Maxwell discusses the way the conformity of molecules to physical standards pointed to the act of a Creator. The fixity of spectral lines established by current spectroscopic measurements indicates that each 'molecule is incapable of growth or decay, of generation or destruction'; the identity of the spectral lines of the same chemical substances observed by the spectroscope in different stars gave evidence that every distant world 'is built up of molecules of the same kinds as those which we find on earth'. Immutability and uniformity imply a standard:

> [every] molecule . . . throughout the universe, bears impressed upon it the stamp of a metric system as distinctly as does the metre of the Archives at Paris, or the royal cubit of the Temple at Karnac.

And this conformity to a physical standard, Maxwell concludes, gives a molecule 'as Sir John Herschel has well said, the essential character of a manufactured article, and precludes the idea of its being eternal and self-existent' (*SP*, **2**: 376). Measurement implies standards, based on the fundamental units of mass, length and time; and the essentially arbitrary standards of physical measurement represent features of nature beyond mere human contrivance.

It is this concern with measurement and the establishment of 'a numerical estimate of some magnitude', that Maxwell states as the true aim of 'Experiments of Research' in his inaugural lecture at Cambridge in October 1871. Such experiments are not primarily concerned 'to see what happens under certain conditions' but rather 'to measure something which we have already seen'. He declared that experiments of this kind, where measurement is involved, 'are the proper work of a Physical Laboratory'. He was careful to emphasise that he did not wish to imply that physics will degenerate into a preoccupation with 'measurement to another place of decimals'. And he reassured his Cambridge audience that the new laboratory would not become 'a place of conscientious labour and consummate skill . . . classed with the other great workshops of our country'. The study of experimental physics would be subsumed within the liberal culture of the university: 'the labour of careful measurement . . . [will be] rewarded by the discovery of new fields of research'. The cultivation of accurate measurements is therefore part of an essentially creative enterprise, 'preparing the materials for the subjugation of new regions' (*SP*, **2**: 242–4). The imperial metaphor helped to associate the study of experimental physics with training for the Indian Civil Service rather than the workshops of Birmingham.

Maxwell was well aware of the resistance that laboratory study, with its associations of craft and workshop expertise, would be likely to encounter within the wrangler culture of Cambridge. The mathematical physics of the *Treatise on Electricity and Magnetism* (1873), which includes gems such as his polar representation of spherical harmonic functions and his treatment of Green's function and Green's reciprocity theorem, was consonant with Cambridge mathematics; but his discussion of instruments and observations was alien. Writing to John William Strutt (later Lord Rayleigh) in March 1871, after his appointment to the Cambridge chair, Maxwell observed that

> it will need a good deal of effort to make Exp. Physics bite into our University system which is so continuous and complete without it.
>
> To wrench the mind from symbols and even from experiments on paper to concrete apparatus is very trying at first, though it is quite possible to get fascinated with a course of observation as soon as we have forgotten the scientific part of it. If we succeed too well, and corrupt the minds of youth, till they observe vibrations and deflexions and become Senior Op.s instead of Wranglers, we may bring the whole University and all the parents about our ears. (*LP*, **2**: 615–16)

As Maxwell knew from personal experience, metrology could have its own fascination. But for a man to fall from the Wrangler class to that of a Senior Optime in the examination for the Mathematical Tripos could have serious implications for his career. In the event, during Maxwell's tenure of the professorship in the 1870s, the new Cavendish Laboratory was the preserve of a handful of graduates, mainly wranglers, not of undergraduates inveigled from preparation for the all-important examination.

When compared with his Aberdeen lecture of 1856, Maxwell's Cambridge inaugural lecture shifts the emphasis from the importance of measurement and quantification in physics, to espousing precise experimentation and exactitude as the culture of laboratory practice, leading to the territorial expansion of science. As an example of how the labour of careful measurement led to the discovery of new fields of research, he instanced the recent history of the study of terrestrial magnetism. Beginning with Alexander von Humboldt's encouragement of international cooperation as a means to obtain accurate measurements of the earth's magnetism, he describes the work of the Magnetic Union, inspired by Carl Friedrich Gauss and Wilhelm Weber in the 1830s. This endeavour, Maxwell remarks, led Weber 'to the numerical determination of all the phenomena of electricity' (*SP*, **2**: 244–6).

In part this shift in emphasis from the Aberdeen inaugural lecture echoes the transformation in Maxwell's own laboratory practice and style of experimentation between 1856 and 1871. In the 1850s he was engaged in studies of colour vision, using apparatus of simple construction, his work shaped by contemporary practice but embodying the values of Cambridge mathematics. In the early 1860s he became involved in a collaborative endeavour to establish a standard of electrical resistance. Maxwell and his colleagues worked with precision apparatus designed by William Thomson, aiming to achieve a high level of accuracy in their measurements in seeking to establish an electrical standard. For Maxwell, this undertaking embodied a major advance in experimental technique and laboratory culture.

This work led him to two major experimental projects which were fundamental to his broader theoretical objectives, and which opened up new fields of research. The first involved the application of the technique employed in the electrical measurements, 'studying the oscillations of magnets by aid of mirrors', to the study of gas viscosity: 'the determination of gaseous friction by means of a disc oscillating in a gas' (*LP*, **2**: 96), as he described it to Stokes in June 1863 at the time of the electrical experiments, when the application of the method occurred to him (see Chapter VIII.2). The second project was the determination of the ratio between the electrostatic and electromagnetic units. According to his theory of electromagnetism, the coincidence between this ratio and the velocity of light established that light was due to transverse waves in the electromagnetic ether (see Chapter V.2). His experimental determination of the ratio of units was a consequence of the need to establish this equivalence on secure foundations. In these two projects Maxwell employed specially constructed precision apparatus, adhering to the rigorous standards of precision and numerical measurements that had transformed the practice of experimental physics around the middle of the nineteenth century.

The work of Gauss and Weber on terrestrial magnetism provided an important exemplar for the programme of accurate experimentation using precision apparatus. Unimpressed with Humboldt's methods, Gauss contrasted current work on the earth's magnetism with the precision instruments used by astronomers. Moreover, he introduced 'absolute' units of measurement, to reduce measures of the earth's magnetic force to the fundamental units of mass, length and time. Gauss' first magnetometer, constructed in 1832, provided measurements of magnetic declination; in 1837 he developed an improved instrument, the bifilar magnetometer (suspended from a loop of steel wire rather than a single thread), which gave precise observations of magnetic intensity. His collaboration with the physicist Wilhelm Weber led to the public dissemination of their instruments, observations and theoretical results.

In his classic experimental and theoretical studies of electrodynamics in the 1840s, Weber used an 'electrodynamometer', an instrument he developed for the precise measurement of electrical forces. This was a delicately constructed instrument in which the rotation of a suspended coil (under the effect of electrical force) was counteracted by the suspension of the coil. The delicate deflections of the coil were designed to provide precise readings to reveal mathematical laws of electrodynamics: its construction required close collaboration between instrument maker and physicist. His paper on 'electrodynamic measurements', published in the *Annalen der Physik* in 1848 and translated in Taylor's *Scientific Memoirs*, established a paradigm for future research.[36]

In characterising the transition to new standards of experimental investigation, Maxwell placed emphasis on the way mathematical analysis helped to shape the construction of appropriate instruments. Writing in 1873, he noted that it was by understanding the mathematical principles of electrostatics that William Thomson had been able to improve the design of the attracted disc electrometer, first constructed by William Snow Harris in the 1830s.

> Sir W. Thomson has improved methods somewhat similar by the help of his mathematical powers till they have attained still greater accuracy, and can estimate the quantities in absolute measure.
>
> (*LP*, **2**: 858–9)

He had corresponded with Thomson in September 1868 on the electrometer, outlining 'a theory of your disk and guard ring opposed to a parallel large plate' (*LP*, **2**: 428). The instrument was described and its theory discussed in the *Treatise on Electricity and Magnetism* (*Treatise*, **1**: 245–6 (§201); **1**: 266–7 (§§216–18)). Mathematical analysis was an essential component of instrument design and use.

The demands of the new technology of cable telegraphy in the 1850s prompted the determination of a standard of electrical resistance.[37] In 1860, Werner Siemens introduced an arbitrary standard based on the resistance of a tube of mercury. But prompted by Weber's advocacy, Thomson successfully urged the British Association Committee on standards of electrical resistance, formed in 1861, to adopt an 'absolute' system based on units of mass, length and time. The aim was to establish units that would meet the needs of physicists and telegraph engineers. Maxwell was co-opted as a member of the Committee in 1862, and in May and June 1863 he joined Fleeming Jenkin (a telegraph engineer) and Balfour Stewart (an experimental physicist) in an accurate measurement of electrical resistance, employing a method devised by Thomson. The experiments were performed at King's College London, where Maxwell was professor of natural philosophy.

In Thomson's apparatus the resistance of a rotating coil was calculated from the measurement of the deflection of a magnet placed at its centre. The experiments required precise measurements of the length of the wire in the coil, and of the time of 100 revolutions of the coil (*LP*, **2**: 98, 103). The apparatus consisted of the driving gear, the revolving coil, a governor (designed by Jenkin) to control its speed, the scale with a telescope by which the deflections of the magnet were observed, and an electric balance by which the resistance of the copper coil was compared with a standard.[38] Copies of the standard resistance issued by the Committee were made available for sale in February 1865 (*LP*, **2**: 214), and the British Association unit of resistance, known as the 'ohm', was used by physicists and cable companies.

In the 'Preface' to his *Treatise on Electricity and Magnetism*, Maxwell noted that the application of the science of electromagnetism to telegraphy had 'reacted on pure science by giving a commercial value to accurate electrical measurements' (*Treatise*, **1**: viii). Shortly afterwards, in a review of Jenkin's text on *Electricity and Magnetism*, he emphasised the importance of the 'testing-office and the engineer's specification', for telegraph engineers were concerned with the science of 'currents and resistance to be measured and calculated' (*LP*, **2**: 842). These concerns were to be fully represented in the work of the Cavendish Laboratory. Shortly following his election to the professorship at Cambridge in March 1871 Maxwell informed Tait that he intended 'to try and settle the Ohm again' (*LP*, **2**: 649), for it had become clear that the King's College experiments were insufficiently accurate, and that there was significant error in the resistance standards issued in 1865. Maxwell had already inquired of Thomson about obtaining 'from the B.A. some of their apparatus for the Standard committee. In particular the spinning coil . . .' (*LP*, **2**: 627), and the equipment was secured for Cambridge. This projected re-determination of the standard 'ohm', though not carried through in his lifetime, became a major objective in Maxwell's direction of the Cavendish Laboratory.[39]

Maxwell's theoretical account of electromagnetism in the *Treatise* commences with a discussion of 'the measurement of quantities':

> in all scientific studies it is of the greatest importance to employ units belonging to a properly defined system . . . by ascertaining the *dimensions* of every unit in terms of the three fundamental units . . . of Length, Time, and Mass . . . . A knowledge of the dimensions of units furnishes a test which ought to be applied to the equations resulting from any lengthened investigation. (*Treatise*, **1**: 1–2)

Maxwell had developed his dimensional analysis in a paper written in collaboration with Fleeming Jenkin, which was included in the Committee's 'Report' in 1863, and reprinted with revisions as a separate paper two years later.[40] Here Maxwell introduced the dimensional notation, which was to become standard, expressing the dimensional relations as products of powers of Mass, Length and Time. For every quantity, the ratio of the two absolute definitions (of the electrostatic unit based on forces between electric charges and the electromagnetic unit based on forces between magnetic poles), is a power of a constant unit with dimensions $[LT^{-1}]$ and magnitude very nearly the velocity of light.

According to Maxwell's electromagnetic theory, as formulated in his paper 'On physical lines of force' in 1862, this ratio between electrostatic and electromagnetic units, symbolised as v, corresponds to the propagation of a disturbance in an electromagnetic ether. The coincidence of this ratio and the velocity of light led him to assert that light was due to transverse waves in the electromagnetic ether (see Chapter V.2). The new dimensional argument established a purely phenomenological link between electromagnetic quantities and the velocity of light. This may have helped to foster the strategy of his paper 'A dynamical theory of the electromagnetic field' (1865), where he sets out a formulation of his theory of the electromagnetic field in which the theory is detached from the mechanical ether model in which it had been initially embedded (see Chapter VI.1).

The study of dimensional relations also revealed the different classes of experiments from which the ratio of electrostatic and electromagnetic units of electricity, and hence the velocity of propagation of electromagnetic waves, could be determined (*LP*, **2**: 110–11). When in autumn 1861 Maxwell had written to inform Michael Faraday and Thomson of his conclusion that 'the magnetic and luminiferous media are identical', he had based this claim on a comparison between Weber and Kohlrausch's experimental value for the ratio of units, v, and various values which had been given for the velocity of light. He declared that the 'coincidence is not merely numerical', and affirmed that he had 'made out the equations in the country before I had any suspicion of the nearness between the two values of the propagation of magnetic effects and that of light' (*LP*, **1**: 685, 695). He made no explicit numerical estimate of the closeness of agreement: the numbers differ by about 1 per cent.[41] A new value for the speed of light established by Léon Foucault in 1862 widened the gap to about 4 per cent.[42] Maxwell acknowledged Foucault's 'more accurate experiments' in his paper 'A dynamical theory of the electromagnetic field' (*SP*, **1**: 580), though without commenting on the implications for his electromagnetic theory of light.

In the autumn and winter of 1864–5 he corresponded with Thomson over various methods of obtaining accurate measurements of v. He outlined a 'plan to weigh an electrostatic attraction against an electromagnetic repulsion directly' (*LP*, **2**: 176), establishing the principle of the method he was to employ. This would, he explained, involve an arrangement 'in which the electric attraction of 2 discs is balanced by the electromagnetic repulsion of two coils' (*LP*, **2**: 211); and in April 1865 he confirmed his adoption of this method.

> The advantage of equilibrating electrostatic attraction with electromagnetic repulsion derived from the same source, instead of balancing each separately by weights or springs is that the forces are applied simulta-

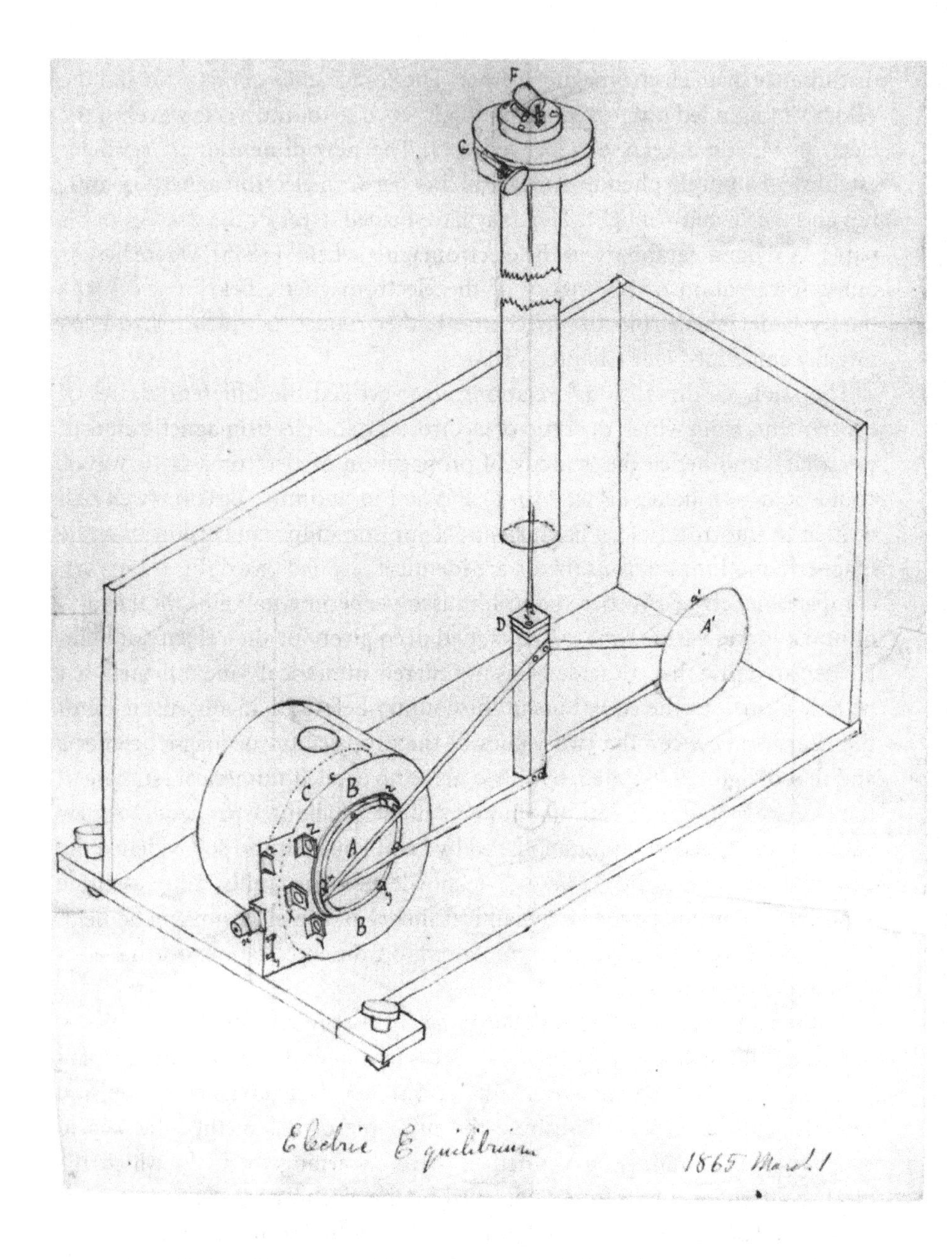

Plate II Maxwell's drawing (1865) of the torsion balance for an experiment for the determination of the ratio of the electrostatic to the electromagnetic unit of electricity.

> neously to a body already in equil^{m} and you have no trouble about unstable eq^{m} when you have eq^{m} at all, and the instability may be just overcome by the elasticity of the suspension. (*LP*, **2**: 219)

On 1 March he had carefully drawn the torsion balance, to be constructed by a London instrument maker, with a counterpoise coil traversed by the same current in the opposite direction, to eliminate the effects of terrestrial magnetism on the suspended coil (*LP*, **2**: 213).

The experiments were finally performed by Maxwell and Charles Hockin (a graduate of the Mathematical Tripos, who had been recruited to assist in the experiments on the 'ohm'), in London in May 1868.

> The experiments consisted in observing the equilibrium of two forces, one of which was the attraction between two disks, kept at a certain difference of potential, and the other was the repulsion between two circular coils, through which a certain current passed in opposite directions. For this purpose one of the disks, with one of the coils attached to its hinder surface, was suspended on one arm of a torsion-balance, while the other disk, with the other coil behind it, was placed at a certain distance, which was measured by a micrometer-screw. The suspended disk, which was smaller than the fixed disk, was adjusted so that in its position of equilibrium its surface was in the same plane with that of a 'guard-ring', as in Sir W. Thomson's electrometers, and its position was observed by means of a microscope directed on a graduated glass scale attached to the disk. In this way its position could be adjusted to the thousandth of an inch, while a motion of much smaller extent was easily detected. (*LP*, **2**: 377)

The experiment was delicate and intricate; for it to be carried to a conclusion required dexterity, patience, and a theoretical grasp of its principles.[43]

In a letter of reference written at the time Maxwell commended Hockin's 'very extensive mathematical knowledge', along with his capacity for 'hunting for sources of discrepancies' and for surmounting 'the physical difficulty of keeping up the spirit of accuracy to the end of a long and disappointing day's work' (*LP*, **2**: 409). The style of experimentation required mathematical expertise joined with a capacity for precise measurement. This was, Maxwell announced in his Cambridge inaugural lecture, to be the style of work in the Cavendish Laboratory.

From these experiments Maxwell obtained a value for *v* which was nearly 4 per cent lower than Foucault's value for the velocity of light (*SP*, **2**: 135).[44] In the *Treatise* his conclusion was expressed in a more cautious and restrained tone than in his heady claims to Faraday and Thomson in late 1861:

> It is manifest that the velocity of light and the ratio of the units are quantities of the same order of magnitude. Neither of them can be said to be determined as yet with such a degree of accuracy as to enable us to assert that the one is greater or less than the other. It is to be hoped that, by further experiment, the relation between the magnitudes of the two quantities may be more accurately determined.
>
> (*Treatise*, **2**: 387–8 (§787))

Shortly after these words appeared in print he reported to the Royal Society on a paper by Dugald M'Kichan (working in Thomson's Glasgow laboratory) on the determination of v. Maxwell had advised M'Kichan on the experimental arrangement, which differed from his own method, and approved his 'investigation as probably far more accurate than any yet made', giving a value for v within 2 per cent of Foucault's value for the velocity of light.[45] These results had the advantage of being based on use of Thomson's new electrometers, but relied on the value of the British Association 'ohm'. Looking to re-determine the 'ohm' in the Cavendish Laboratory, Maxwell noted that improvements in determining v would depend upon any 'correction hereafter discovered to be necessary in the received value of this unit' (*LP*, **2**: 849–51).

The establishment of the professorship of experimental physics at Cambridge was linked to demands for reform of the Mathematical Tripos by the inclusion of physical subjects into the examination. This process of reform of the Tripos was gently fostered by Maxwell; on his appointment as an examiner and moderator four times between 1866 and 1870, he introduced a few questions on electricity, magnetism and heat into the examination. Reporting in February 1869, a Physical Sciences Syndicate recommended the establishment of a new professorship to offer lectures in these subjects, and also urged the foundation of a physical laboratory. The University's scientific professors had already concluded that a new professorship would be required to provide effective teaching in the new subjects, the major fields of research in mathematical physics since the 1840s. These recommendations had uncomfortable financial implications; but the Chancellor of the University, the Duke of Devonshire, intervened with an offer 'to provide the funds required for the building and apparatus, as soon as the University shall have in other respects completed its arrangements for teaching Experimental Physics, and shall have approved the plan of the building'. The new professorship of experimental physics was advertised on 14 February 1871, the election being announced for 8 March.[46]

Urging Maxwell to stand for election (Thomson having declined), Stokes explained that his lectures would be 'subject to the approval of the board of Mathematical Studies'; it was anticipated that candidates for the Mathematical Tripos, as well as candidates for the Natural Sciences Tripos, would attend.[47] In fulfilling the requirement that his professorship meet the scope of the Mathematical Tripos when the new regulations came into force in 1873, Maxwell lectured on 'Heat' in Michaelmas term 1871, on 'Electrostatics and Elektrokinematics' in Lent term 1872, and on 'Electromagnetism' in Easter term 1872, the titles of these courses being modified slightly in subsequent years. With very few exceptions, the students attending in the early years were candidates for the Mathematical Tripos.[48]

Maxwell's duties at Cambridge thus fell under two headings: to provide instruction in the mathematical physics of heat, electricity and magnetism for the Mathematical Tripos, and to design and direct the new physical laboratory. He aimed to turn the energies of the mathematicians towards experimental physics, to 'corrupt the minds of youth, till they observe vibrations and deflexions', their minds wrenched from symbols (*LP*, **1**: 615–16). Maxwell thus wrote in jocular vein to John William Strutt in March 1871, having accepted the professorship, in reply to a letter from Strutt urging him to stand for election. 'What is wanted by most who know anything about it', Strutt had written, 'is . . . a mathematician who has actual experience in experimenting, & who might direct the energies of the younger Fellows & bachelors into a proper channel'.[49]

Strutt, senior wrangler in 1865 and a Fellow of Trinity, already following the path of a mathematical and experimental physicist, and who was to be Maxwell's successor in the professorship, rightly perceived the role of the new professor and laboratory within the educational structure of Cambridge. Maxwell reported his initial efforts, in instructing his new students in taking galvanometer readings using the British Association coils, in letters to the American physicist Henry Augustus Rowland in July 1874 and to Fleeming Jenkin the following November. Distinguishing three different deflections of the galvanometer needle, as the 'shove' of the coils on the needle, the 'kick' which measured a transient current, and the 'jerk' which measured the time-integral of this current, he translated the observations of the experimenters into mathematical language. He expressed the point succintly in his letter to Jenkin.

We are now going ahead with the comparison of coils and several men are able to rig up the various combinations out of their own heads, now that they know by inspection the difference between a

shove $\propto x$ (x = current)
kick $\propto \int x \, dt$
jerk $\propto \iint x \, dt \, dt$.[50]

Maxwell's task was to turn wranglers into physicists.

IV Physical and geometrical analogy

IV.1 The language of field theory: Faraday and Thomson

Writing to William Thomson in September 1855, Maxwell reported his progress with work on which he had been engaged for about 18 months:

> Now I have been planning and partly executing a system of propositions about lines of force &c which may be *afterwards* applied to Electricity, Heat or Magnetism or Galvanism [Electromagnetism], but which is in itself a collection of purely geometrical truths embodied in geometrical conceptions of lines, surfaces &c.
>
> (*LP*, **1**: 320)

He was alluding to his first paper on the theory of the electromagnetic field, 'On Faraday's lines of force' (1856), which he was shortly to present to the Cambridge Philosophical Society, in December 1855 and February 1856.

In this paper Maxwell formulates laws expressing relations between magnetic forces and electric currents; these laws are grounded on a geometrical imagery of interacting curves and lines passing through surfaces. To represent the '*direction* of the force' he deploys Faraday's theory of lines of force, providing a 'geometrical model of the physical phenomenon'. But to represent the '*intensity* of the force at any point' he introduces a 'physical analogy', the 'geometrical idea of the motion of an imaginary fluid'. He had adopted this method, based on geometry and the 'resemblance in mathematical form' between electricity and the motion of a fluid, rather than espouse a 'physical hypothesis' in which 'physical facts will be physically explained', because of the 'rashness in assumption which a partial explanation encourages'. He wishes 'to obtain physical ideas without adopting a physical theory' (*SP*, **1**: 155–9). This attempt to explicate a physical geometry of the electromagnetic field is based on geometry and analogy, a style of argument which he contrasts with an appeal to hypotheses and models, as later deployed in his paper 'On physical lines of force' of 1861–2 (see Chapter V.2). These are important themes in Maxwell's natural philosophy.

He had announced his intention to pursue this subject in February 1854, after his graduation at Cambridge, telling Thomson that he intended 'to return to Physical Subjects . . . to attack Electricity' (*LP*, **1**: 237). Given the scope of his interests before going up to Cambridge (see Chapter II.1), his decision to address a physical topic, rather than continue his current work on the geometry of surfaces (see Chapter II.2), was not surprising. Electricity, though at the time excluded from the Mathematical Tripos, was open to incorporation within Cambridge 'mixed mathematics'. The contrast between 'geometrical truths' and a 'physical hypothesis' evoked the preference, characteristic of the work of Stokes and Thomson at the time, for a mathematical theory rather than a physical model, a distinction Maxwell had deployed in his paper 'On the equilibrium of elastic solids' (1850).

Maxwell had been acquainted with Thomson since 1850, if not earlier (*LP*, **1**: 205), and had followed his work on the theory of electromagnetism. The title of Maxwell's brief manuscript on the 'Mathematical Theory of Polar Forces' (*LP*, **1**: 210–11) is drawn from an expression in Thomson's 1851 paper 'A mathematical theory of magnetism',[1] which it clearly post-dates. In this manuscript Maxwell is concerned with Thomson's application of the 'Potential Function' to the mathematical theory of magnetism.

The concept of potential, which had been developed by Laplace in the theory of gravitation, and subsequently applied to electrostatics by Poisson and Green, was to play an important role in Maxwell's mathematical representation of the physical concept of the field. In this mathematical method the potential is a function which measures the force exerted at a point in space by a mass or charge. In Poisson's equation for electric charge, the distribution of charge is related to the potential in a partial differential equation. The concept of the potential came to have a crucial role in the mathematical theories of the ether and the electromagnetic field, because the partial differential equation for the potential provides an expression for a theory of continuous action in a medium. Thus Maxwell later explained that he considered the concept of the potential to be appropriate to the representation of Faraday's idea of lines of force traversing space, providing the basis for his theory of the field, in which contiguous elements of the field transmit force and energy (*Treatise*, **1**: xi, 99 (§95)).

In the manuscript on the 'Mathematical Theory of Polar Forces' Maxwell advances from an interpretation of the potential function as a 'mathematical abstraction', as a means of representing gravitational and electrostatic attraction (in the work of Laplace, Poisson and Green), to its conceptualisation as a physical property of space.

> We have no reason to believe that anything answering to this function has a physical existence in the various parts of space, but it contributes not a little to the clearness of our conceptions to direct our attention to the potential function as if it were a real property of the space in which it exists.

He illustrates this interpretation by pointing to the duality of forces in electricity and magnetism; Thomson had made reference to Faraday's analogy between states of magnetic and electric polarity in magnets and dielectrics. Maxwell observes that this 'opposition of properties in opposite directions constitutes the polarity of the element of space' (*LP*, **1**: 210–11). Given this prior interest in Thomson's application of potential theory to the physical ideas articulated by Faraday, it was natural for him to write to Thomson for advice on a course of 'reading & working' (*LP*, **1**: 237).

While it seems likely that Maxwell's interest in the theory of electricity and magnetism had been sparked by Thomson's application of potential theory, it was Faraday's work, as he recollected in the 'Preface' to his *Treatise on Electricity and Magnetism* (1873), to which he had first turned: 'before I began the study of electricity I resolved to read no mathematics on the subject till I had first read through Faraday's *Experimental Researches in Electricity*'. He claimed that his own contribution had been to have 'translated what I considered to be Faraday's ideas into a mathematical form'. He recalled that 'As I proceeded with the study of Faraday, I perceived that his method of conceiving the phenomenon was also a mathematical one, though not exhibited in the conventional form of symbols'. In illustration, he pointed to Faraday having envisioned 'lines of force traversing all space'; and also noted Faraday's emphasis on a 'medium' as the embodiment of the electromagnetic field. In Maxwell's view these concepts 'were capable of being expressed in the ordinary mathematical forms' (*Treatise*, **1**: ix–x).

He explained that the crucial concept in this process of mathematical translation was the 'theory . . . of the potential', which 'belongs essentially to the method which I have called that of Faraday'. Thomson was his main guide in the provision of appropriate theorems in potential theory, in their application to Faraday's concept of 'lines of force' and supposition of 'real actions going on in the medium' between electrified bodies. In Maxwell's view these distinctively Faradayan concepts were intrinsically mathematical. Diagrams of lines of force cutting equipotential surfaces could be 'studied as illustrations of the language of Faraday in speaking of "lines of force"'. The concept of potential 'considered as a quantity which satisfies a certain partial differential equation' was 'the appropriate expression for a theory of action exerted between contiguous parts of a medium', and therefore expressed the essentials of Faraday's attempt to find 'the seat of the phenomena in real actions going on in the medium' (*Treatise*, **1**: x–xi, 99, 146 (§§95, 122)).

Maxwell made the Faradayan concepts of lines of force and action in an ambient medium central to his theory of the electromagnetic field. He incorporated these two modes of representation of the field into his first two papers on field theory. In 'On Faraday's lines of force' (1856) he developed Faraday's theory of the primacy of lines of force; while in his paper 'On physical lines of force' (1861–2) he proposed a mechanical illustration of Faraday's theory of electrostatic action between the contiguous particles of the dielectric medium, a model that was fundamental to the discovery of his electromagnetic theory of light (see Chapter V.2). When he highlighted Faraday's two seminal concepts in the *Treatise*, he was pointing to the two modes of representation of the electromagnetic field that had shaped his own work.[2]

Maxwell's declaration, in an 1873 portrait of Faraday, that the theory of lines of force showed Faraday to be 'in reality a mathematician of a very high order' (*SP*, **2**: 360), points to an important feature of his espousal and interpretation of Faradayan ideas. In his paper 'On Faraday's lines of force', Maxwell was concerned to defend Faraday from the charge that his 'processes of reasoning . . . are of an indefinite and unmathematical character'. In this paper he aimed to refute this dismissive opinion, interpreting the notion of lines of force by a 'geometrical model' of lines cutting surfaces, and appealing to the 'resemblance in mathematical form' between electricity and the 'physical analogy' of the flow of an incompressible fluid (*SP*, **1**: 156–8). These propositions could moreover be expressed in an analytical form when this was judged to be appropriate. The physically intuitive language of Faraday was therefore mathematical in character.

Faraday's extraordinary series of experimental discoveries, of electromagnetic induction, the laws of electrochemistry, and magneto-optic rotation (the Faraday effect), formed 'the nucleus of everything electric since 1830', Maxwell told R. B. Litchfield in 1858 (*LP*, **1**: 582). Faraday's most famous concept was that of magnetic curves or lines of force. He explained that by 'magnetic curves I mean the lines of magnetic forces . . . which would be depicted by iron filings'. On discovering electromagnetic induction in 1831, he had explained the induction of an electric current between the primary and secondary circuits wrapped round an iron ring in terms of the creation of a state of electrical 'tension' in the secondary wire, an electrical condition of matter which he termed 'the *electro-tonic* state'. The creation and dissolution of the electro-tonic state gave rise to electromagnetic induction. Faraday conceived these two concepts, lines of force and the electro-tonic state, as alternative modes of representation. As a result of experiments on inducing currents by the rotation of magnets, he concluded that 'a *singular independence* of the magnetism and the bar in which it resides is rendered evident'. In consequence, he rejected his explanation of electromagnetic induction in terms of 'that peculiar condition [of matter] . . . the electro-tonic state', in favour of the suggestion that the 'magnetic curves' of the primary circuit were cut by a current-carrying wire of the secondary circuit.[3]

In studying electrolysis and electrostatic induction in the 1830s, Faraday emphasised the transmission of forces mediated by particles of matter in the space between electrically charged bodies. In his investigation of electrostatic induction he pointed to the role of the ambient medium, the 'dielectric'. He found that different dielectrics had different capacities, termed their 'specific inductive capacity', for the mediation of electrostatic forces, and that induction took place in 'curved lines'. Electrostatic induction was analogous to magnetism rather than to gravity: 'the whole action is one of contiguous particles, related to each other, not merely in the lines which they may be conceived to form through the dielectric . . . but in other lateral directions also'. He explained that by 'contiguous particles' he meant 'those which are next to each other, not that there is *no* space between them'.

Electrostatic induction was analogous to electrolysis, where Faraday had supposed that 'the whole effect in the electrolyte appeared to be an action of the particles thrown into a peculiar or polarized state'. He concluded that electrostatic induction 'never occurred except through the influence of the intervening matter'. He recognised that his theory contradicted the received interpretation of electrostatics, where induction was conceived 'as an action at a distance and in straight lines'. In his view, by contrast, the particles of the dielectric mediated the action, being 'thrown into a state of polarity and tension', a 'polarized state', the 'particles assuming positive and negative points or parts'; this was a 'forced state' of matter.[4] This theory of electrostatics was generally thought anomalous, incompatible with the mathematical theory based upon direct action at a distance. But in 1845, in the first of a series of papers on the application of potential theory to electrostatics,

Thomson established that Faraday's theory and conventional mathematical representations could be reconciled.

Prompted by the important new studies of magnetism which he commenced in 1845, Faraday's notion of particles mediating forces yielded to an emphasis on the primacy of lines of force. That year, he discovered the Faraday effect, the rotation of the plane of polarisation of linearly polarised light transmitted through a 'diamagnetic' in a 'magnetic field' (a term Faraday here used for the first time). His theory of the 'magnetic field' sought to explain magnetic and electromagnetic phenomena in terms of the disposition of lines of force in space: paramagnetism and diamagnetism were explained by the tendency of substances to conduct lines of force relative to the ambient magnetic field. In a paper 'On the physical character of the lines of magnetic force' (1852) he affirmed the physical reality of the lines of force, illustrated by the patterns assumed by iron filings sprinkled over magnets. He declared that

> the outer forces at the poles can only have relation to each other by *curved* lines of force through the surrounding space; and I cannot conceive curved lines of force without the conditions of a physical existence in that intermediate space.

The lines of force described the physical existence and reality of the magnetic field: the role of mediation was now ascribed to lines of force.[5]

In this paper Faraday introduced an image to represent 'the well-known relation of the electric and magnetic forces'. The two 'axes of power', electric and magnetic, were represented by two rings perpendicular to each other. One ring represents a current of electricity, which generates lines of magnetic force in the other ring. He pointed to 'the intimate physical relation . . . the oneness of condition of that which is apparently two powers or forms of power, electric and magnetic'. This image of interacting rings gave expression to his conception of the physical field. The 'mutual relation of the magnetic lines of force and the electric axis of power' implied that magnetic lines of force possess a 'dynamic condition' analogous to electric currents. In Maxwell's conceptualisation in 'On Faraday's lines of force', this dynamic power was represented by the 'intensity' of force summed along lines of force (*SP*, **1**: 206–7).

To represent this dynamic condition, using terms which Maxwell was to seize upon, Faraday reintroduced the concept of the electro-tonic state, 'to constitute and give a physical existence to the lines of magnetic force'. He explained that

> Again and again the idea of an *electro-tonic* state . . . has been forced on

> my mind; such a state would then constitute the physical lines of magnetic force.

The image of perpendicular rings of electric and magnetic lines as interacting curves or axes of power expressed his concept of the physical field; electric currents interacted with the electro-tonic state.[6] The lines of force could also be given quantitative measure: 'the relative amount of force, or of lines of force in a given space, [is] indicated by their concentration or separation, i.e. by their number in that space'.[7] To expound Faraday's theory Maxwell introduced the concept of 'quantity' to denote force summed across lines of force in a given space (*SP*, **1**: 206–7). Thus as he remarked in 1873, Faraday's theory of lines of force provided 'a complete definition of the magnetic state of every part of the field' (*LP*, **2**: 805), in a form amenable to expression in Maxwell's mathematical language of 'intensity' and 'quantity', forces summed along and across lines of force.

Maxwell made rapid progress in his study of electricity and magnetism. Writing to Thomson in November 1854, having inquired about a course of reading only in February that year, he was able to outline theorems in electromagnetism. These theorems, which he reformulated in his paper 'On Faraday's lines of force', express his interpretation of Faraday's concept of the interaction of electric currents and magnetic lines of force (see Chapter IV.2). He remarked to Thomson that 'I have heard you speak of "magnetic lines of force" & Faraday seems to make great use of them'. He went on to signal his assimilation of Faraday's theory of the 'magnetic field' as formulated in 1852:

> Now I thought that as every current generated magnetic lines & was acted on in a manner determined by the lines thro wh: it passed that something might be done by considering . . . [the] 'magnetic field' or space and developing the geometrical ideas according to this view.

Faraday's concept of interacting curves was inherently geometrical, which facilitated its intelligibility. But in developing his geometrical representation of 'magnetic lines of force' and the 'magnetic field', Maxwell told Thomson, he was indebted to Thomson's discussion of the mathematical relation between attraction in electrostatics and the conduction of heat: 'I was greatly aided by the analogy of the conduction of heat, wh: I believe is your invention at least I never found it elsewhere' (*LP*, **1**: 254–5).

Maxwell was alluding to Thomson's analogy, proposed in a major paper published in 1842, between the flux of heat across isothermal surfaces and electrostatic attraction across equipotential surfaces. The analogy suggested links in mathematical form between the conduction of heat, electrostatics, and the flow of a fluid. This application of potential theory formed the basis of Maxwell's statement of the method of 'physical analogy' in 'On Faraday's lines of force': 'that partial similarity between the laws of one science and those of another which makes each of them illustrate the other' (*SP*, **1**: 156).

Thomson's analogy between electrostatics and the steady flow of heat developed from his work on Fourier's theory of heat. The analogy is between temperature and potential (though Thomson did not use this term), and he expounded it using geometrical imagery. The flux of heat and electrostatic attraction flow continuously across isothermal and equipotential ellipsoidal surfaces; the curves representing the flux of heat and attraction cut 'all the surfaces perpendicularly'. Propositions in heat conduction could be translated into theorems in electrostatics:

> if a surface in an infinite solid be retained at a constant temperature, and if a conducting body, bounded by a similar surface, be electrified, the flux of heat, at any point, in the first case, will be proportional to the attraction on an electrical point similarly situated in the second; and the direction of the flux will correspond to that of the attraction.[8]

Shortly afterwards Thomson accepted Gauss' recent use of the term 'potential' to represent electrical attraction; and in 1845 he discovered Green's prior and identical usage of this term,[9] and was able to reformulate his argument accordingly:

> The problem of *distributing sources of heat*... is mathematically identical with the problem of distributing *electricity in equilibrium*.... In the case of heat, the *permanent temperature* at any point replaces the *potential* at the corresponding point in the electrical system, and consequently the *resultant flux of heat* replaces the *resultant attraction* of the electrified bodies, in direction and magnitude.[10]

These remarks appear in the paper in which Thomson first proved that Faraday's theory of electrostatics as mediated by a dielectric was compatible with the current mathematical theory based on action at a distance. He came to realise that Faraday's image of the transmission of forces between the contiguous particles of a dielectric could be interpreted as analogous to Fourier's concept of the propagation of heat between particles.[11] Faraday's ideas could therefore be translated into the language of potential theory. But Thomson was careful to limit this relation to a mathematical analogy:

> It is, no doubt, possible that such [electrical] forces at a distance may be discovered to be produced entirely by the action of contiguous

> particles of some intervening medium, and we have an analogy for this in the case of heat, where certain effects which follow the same laws are undoubtedly propagated from particle to particle. . . . We know nothing, however, of the molecular action by which such effects could be produced . . . .[12]

Thomson's distinction between mathematical and physical theorising is echoed by Maxwell. In a draft of 'On Faraday's lines of force' he notes that in the case of the conduction of heat and attraction 'we have a mathematical analogy between two sets of phenomena confessedly very different' (*LP*, **1**: 356). He develops this argument in the published text of the paper. The analogy between heat and attraction is based only on the 'mathematical resemblance of some of their laws', he points out, but this analogy 'may still be made useful in exciting appropriate mathematical ideas'. Hence he will formulate a 'geometrical model of the physical phenomena', a 'physical analogy' rather than a 'physical hypothesis' (*SP*, **1**: 155–8).

In a paper of 1843 Thomson had established the equivalence between considering forces acting at a distance on electrified bodies, or in terms of the action of the atmosphere on the electrified surfaces.[13] On reviewing Thomson's *Reprint of Papers on Electrostatics and Magnetism* (1872), Maxwell observed that, in considering the force in terms of the diminution of atmospheric pressure on the electrified surface, Thomson had employed 'only another name for . . . *tension* along the lines of electric force'. The paper, he declared, may therefore be regarded 'as the germ of that course of speculation by which Maxwell has gradually developed the mathematical significance of Faraday's idea of the physical action of the lines of force' (*SP*, **2**: 304). This remark may well indicate his interpretation of Thomson's paper in 1854, and help to explain the genesis of his rendition of Faraday's 'dynamic condition' of magnetic lines of force as an 'intensity' of forces summed along the lines of force. But this comment reflects Maxwell's own interpretation of the paper, or his subsequent reconstruction of its import, rather than Thomson's viewpoint when writing it, for Faraday's theory of lines of force was not in Thomson's sights in 1843.

But by 1847, in his paper 'On a mechanical representation of electric, magnetic, and galvanic [electromagnetic] forces',[14] Thomson did have Faraday's ideas in mind. This paper, as Maxwell described it in writing to Thomson in September 1855, was based on 'an allegory about incompressible solids' (*LP*, **1**: 322). The allegory was mechanical, in that it represented electric and magnetic forces in terms of the linear and rotational strain of an elastic solid; and Thomson noted the consistency of his model with Faraday's recent discovery of the magneto-optic rotation. But despite his laconic allusions to physical implications, the argument was confined to elaborating mathematical expressions. In 1856 Thomson was to advance significantly in providing a physical theory of magnetism and the magneto-optic effect, in terms of the rotation of elements of the ether: this work was to profoundly shape the development of Maxwell's field theory after 1857 (see Chapter V.2).

Thomson's interest in potential theory led him to publish, in collaboration with his friend Stokes, a series of 'Notes on hydrodynamics' (1847–9),[15] suggesting links between the conduction of heat, electrostatics and the motion of fluids, an analogy he broadened to include magnetism. In his major paper 'A mathematical theory of magnetism' (1851), Thomson conceived the distribution of magnetic force in two different geometrical modes. The first considered a magnet to be divided along lines of magnetisation into tubes, or 'solenoids', this term being derived from Ampère. The second mode considered a magnet divided perpendicular to lines of magnetisation into 'magnetic shells', this being termed a 'lamellar' distribution of magnetism.[16] By November 1854, Maxwell had begun to reformulate this distinction into a fundamental new conceptualisation, based on his rendition of Faraday's theory of the field in terms of forces of 'intensity' and 'quantity' summed along and across lines of force.

The equation of continuity had been Thomson's subject in the first of the 'Notes on hydrodynamics', and in 'A mathematical theory of magnetism' he noted the 'analogy', as he termed it, between the expression for the 'solenoidal' distribution of magnetism and that for the ' "equation of continuity" to which the motion of a homogeneous incompressible fluid is subject'.[17] In the spring of 1855, Maxwell began to review Thomson's discussion of the equation of continuity, going on to formulate theorems on the stability of fluid motion (*LP*, **1**: 291–3, 295–9), and reporting his ideas in detail to Thomson (*LP*, **1**: 309–13). He subsequently began to develop the analogy between magnetic lines of force and streamlines and tubes of flow in an incompressible fluid. This 'physical analogy', the basis of the 'geometrical idea of the motion of an imaginary fluid' which shapes the physical geometry of 'On Faraday's lines of force', was introduced at a relatively late stage in the genesis of the paper. Like other mathematical strands of the argument, it has its origin in Thomson's mathematical theorems, work that Maxwell was to transform into a coherent theory of 'Faraday's lines of force'.

IV.2 Physical analogy and field theory

In September 1855, shortly before presenting his paper 'On Faraday's lines of force' to the Cambridge Philosophical Society, Maxwell wrote to Thomson to indicate the scope of his evolving theory of the electromagnetic field.

> I would be much assisted by your telling me whether you have not the whole draught of the thing lying in loose papers and neglected only till you have worked out Heat or got a little spare time.

Alluding to Thomson's current work on thermodynamics, he goes on to list the reasons which led him to suppose that Thomson had already formulated a systematic field theory of electricity and magnetism.

> That you are acquainted with Faradays theory of lines of force & . . . had the advantage of being well acquainted with *V* [potential] and with Green's essay, and you published a fragment of your speculations in the form of an allegory about incompressible elastic solids.

The 'fragment' Maxwell mentions is Thomson's 1847 'mechanical representation of electric, magnetic, and galvanic forces'; but his analogy to forces in an elastic solid, though suggestive of the propagation of electromagnetic forces by mechanical processes in the ether, was limited to the bare statement of mathematical expressions.

Maxwell goes on to highlight what was for him the most significant of Thomson's contributions to the subject, the distinction between 'solenoidal' and 'lamellar' distributions of magnetism, which, in the second part of 'On Faraday's lines of force', he was to reformulate in providing a mathematical representation of Faraday's theory of lines of force.

> You . . . used a method in your demonstration about the superficial tangential distribution of magnetism in a solenoidal magnet which seems to me to be part of my results applied to magnetism without acknowledging that you had taken it from a more general theory.

All this led Maxwell to surmise that 'you have the mathematical part of the theory in your desk', and that therefore 'all that you have to do is to explain your results with reference to electricity' (*LP*, **1**: 322–3).

This is the nub of Maxwell's relation to Thomson here. Thomson was familiar with Faraday's theory of lines of force; he had developed theorems in potential theory which were applicable to the mathematical expression of Faraday's ideas; and had formulated a mathematical representation of magnetic force, which Maxwell perceived could be developed to give mathematical form to Faraday's theory of lines of force and the 'electro-tonic state'.

These mathematical elements were original to Thomson; but while Maxwell, standing upon Thomson's shoulders, was now in a position to develop a 'more general theory', Thomson was not.

It would be facile to construe their relationship as Thomson the mathematical innovator, Maxwell the physical theorist. While Thomson did not develop his work on electrostatics and magnetism into a rounded and systematic theory, this work was shaped by a theoretical programme.[18] But there is a marked difference in the style of their writings. Many of Thomson's papers are presented as bare-bones mathematics, their physical implications being merely hinted, and their place within a theoretical outlook left obscure. Maxwell, by contrast, wrote as a natural philosopher, always concerned to develop a worldview. As his sequence of memoirs on the electromagnetic field proceeded, he made clear the trajectory of his evolving theoretical vision, from the physical geometry of 'On Faraday's lines of force' to the physical mechanics of 'On physical lines of force' and the analytical mechanics of 'A dynamical theory of the electromagnetic field' (see Chapter I). Yet each of these papers, taken individually and in its own terms, can be read as a unified and systematic explication of the physical field, each memoir based on a different theoretical vantage point. To use Maxwell's own metaphor, 'the focussing glass of theory [can be screwed] . . . sometimes to one pitch of definition, and sometimes to another, so as to see down into different depths' (*LP*, **1**: 377).

Maxwell's achievement over Thomson is in conceptual development, in the creation of a systematic, mathematical physics of the physical field. In so doing he stretches Thomson's mathematics, transforming its meaning. His theory of the field is based on Faraday's imagery (as he describes it in the paper) of '*mutually embracing* curves' (*SP*, **1**: 194n), representing the interaction of lines of force and electric circuits. He achieves his mathematical translation of Faradayan concepts by reformulating Thomson's distinction between 'solenoidal' and 'lamellar' distributions of magnetism, in terms of the intensity–quantity distinction suggested by Faraday's theory of lines of force. In this process of conceptual development he transforms Thomson's mathematical expressions, giving them a meaning foreign to Thomson's original conceptualisation.

Maxwell's remark to Thomson, that 'there can be no doubt that you have the mathematical part of the theory in your desk', served to smooth any ruffled feathers, affirming that he recognised Thomson's contribution in working out the basic mathematical formalism. In similar vein, remarking to Thomson that his expression for the solenoidal distribution of magnetism seemed to have been stated 'without acknowledging that you had taken it from a more general theory', he generously implied that Thomson shared his own theoretical grasp. But Thomson did not have any 'draught of the thing lying in loose papers', in the sense of a systematic theory analogous to and pre-dating Maxwell's theory of Faraday's lines of force.

In placing emphasis on conceptual development, Maxwell did not renounce all claim to mathematical innovation. He concludes 'On Faraday's lines of force' by repeating, publicly and emphatically, his acknowledgement of the value of Thomson's 'mathematical expressions' in having shaped the course of his own work. But the innovations of his argument have a mathematical core; and his conceptual transformation of Thomson's idioms led to the introduction of 'electro-tonic functions', by which he interpreted Faraday's theory of lines of force.

> I may state that the recognition of certain mathematical functions as expressing the 'electro-tonic state' of Faraday, and the use of them in determining electro-dynamic potentials and electro-motive forces is, as far as I am aware, original. (*SP*, **1**: 209)

Maxwell recognised Thomson's rights over his intellectual property. But he notified Thomson, in friendly spirit, of his intention to 'poach', joking that he did 'not know the Game laws & Patent laws of science'; and 'as for the hints you have dropped about the "higher electricity", I intend to take them' (*LP*, **1**: 322–3).

Writing to Thomson in November 1854, Maxwell signalled his adoption of the concept of the 'magnetic field' as conceived by Faraday.

> Now I thought that as every current generated magnetic lines & was acted on in a manner determined by the lines thro wh: it passed that something might be done by considering 'magnetic polarization' as a property of a 'magnetic field' or space and developing the geometrical ideas according to this view.

He describes his proposed representation as 'geometrical' in character, clearly alluding to the relations between lines of force and the space through which they pass, the 'magnetic field'. The concept of 'polarization' is described in phenomenological terms, to represent the attraction of magnets: 'to express the fact that at a point of space the south pole of a small magnet is attracted in a certain direction with a certain force' (*LP*, **1**: 255–6).

To represent the 'magnetic field' Maxwell deploys the concept of 'polarization' in a formal, mathematical sense; and he does so using the imagery of lines of force and the concept of potential. In his manuscript on the 'Mathematical Theory of Polar Forces', in the context of Thomson's reference to Faraday's ideas, he had conceptualised the potential function 'as if it were a real property of the space in which it exists' (*LP*, **1**: 211). He now explains the meaning of 'polarization' in terms of lines of force cutting 'equipotential surfaces (or surfaces perp. to the lines of force)', so that 'space is now cut up into elements'. 'Polarization' is a dual concept, measured along lines of force and across the equipotential surfaces cut by the lines of force.

(1) The pol^{n} of a surface is expressed by the *number* of lines wh cut it.
(2) The pol^{n} of a line is expressed by the number of equip^{l} surfaces wh: it cuts. (*LP*, **1**: 258)

'Polarization' is therefore Maxwell's representation of Faraday's concepts of a 'dynamic condition' along lines of force and the 'concentration' or number of lines of force summed across a surface. He states two 'theorems' to give formal expression to Faraday's concept of magnetic lines of force and electric currents as interacting axes of power.

(1) The pol^{n} of any closed curve is measured by the sum of the intensities of all the currents which pass thro' it.
(2) The pol^{n} of any surface, round the boundary of wh: a current passes, is measured by the intensity of that current. (*LP*, **1**: 257)

The imagery is geometrical, but its formal expression, in terms of lines of force cutting equipotential surfaces, is grounded on the concept of potential. The 'theorems' express symmetrical relations between 'polarization' (in its two senses) and electric currents, expressing Faraday's image of '*mutually embracing* curves' (*SP*, **1**: 194n).[19]

Maxwell's mathematical translation of Faraday's concept of lines of force, in the second part of 'On Faraday's lines of force', has as its keystone a major conceptual transformation: his development of Thomson's distinction between 'solenoidal' and 'lamellar' distributions of magnetism into a distinction between 'magnetic intensity' and 'magnetic induction'. He reformulates Thomson's distinction between two distributions of magnetism, into a distinction between forces summed along lines of force (which he terms 'intensity') and across lines of force ('quantity'). This intensity–quantity dualism had its origins in Faraday's concept of a 'dynamic condition' along lines of force (intensity) and the 'concentration' or number of lines of force summed across a surface (quantity). The distinction between Thomson's two distributions of magnetism, 'solenoidal' (measured along tubes) and 'lamellar' (measured across shells), is reformulated into a distinction between 'magnetic intensity' and 'magnetic induction' (quantity). [In the terminology Maxwell uses in the *Treatise*, the distinction is between 'magnetic force' **H** and 'magnetic induction' **B**.]

In his letter to Thomson of September 1855, Maxwell explains his intention to formulate a mathematical theory of electromagnetism, to encompass

> the theory of the connections of the three divisions of the subject the passage of statical into current electricity and the magnetic properties of closed currents with the laws of induced current. (*LP*, **1**: 321)

He achieves this in the second part of the published paper by introducing his distinction between 'magnetic intensity' and 'magnetic induction', and thus transforming the two 'theorems' of the letter of November 1854 into two 'laws' of electromagnetism. The first 'theorem' was readily transformed:

> *The entire magnetic intensity round the boundary of any surface measures the quantity of electric current which passes through that surface.*
> (*SP*, **1**: 206)

The second theorem, as a reciprocal relation for magnetic induction through a surface, was more problematic.[20] He required an electric 'intensity', a variable 'bearing the same relation to magnetic quantity that magnetic intensity bears to electric quantity', so as to yield 'an expression for the quantity of magnetic induction passing through a closed circuit in terms of quantities depending on the circuit itself, and not on the enclosed space'. Following Faraday's statement that lines of force could be represented by the 'electro-tonic state', Maxwell introduces 'electro-tonic functions' which, he declares, have 'at least a mathematical significance'. The supposition of the 'electro-tonic intensity' [which in the *Treatise* he terms the 'vector-potential of magnetic induction' **A** (see Chapter VII.1)] is justified by Faraday's discovery of the induction of electric currents 'by *changes* of the electric or magnetic phenomena', which led him to the 'conception of a state into which all bodies are thrown by the presence of magnets and currents'. Faraday had 'hinted at the probability that some phenomena might be discovered which would render the electro-tonic state an object of legitimate [investigation]', but Maxwell concedes that there was no 'experimental data for the direct proof of the unknown state'. Thus he notes that this 'representation involves no physical theory, it is only a kind of artificial notation' (*SP*, **1**: 188–9, 205–6; *LP*, **1**: 374).

In his letter of September 1855 he had told Thomson that

> I intend next to apply . . . Faradays notion of an *electrotonic* state. I have worked a good deal of mathematical material out of this vein and I believe I have got hold of several truths which will find mathematical

> expression in the electrotonic state.
> One thing at least it succeeds in, it reduces to one principle not only the attraction of currents & the induction of currents but also the attraction of electrified bodies without any new assumption.
>
> (*LP*, **1**: 322)

The introduction of the 'electro-tonic intensity' as an analogue for 'magnetic intensity' opened the way to the reformulation of the second 'theorem' of the letter of November 1854.

> *The entire electro-tonic intensity round the boundary of an element of surface measures the quantity of magnetic induction which passes through that surface or, in other words, the number of lines of magnetic force which pass through that surface.*
>
> (*SP*, **1**: 206)

The two 'theorems' of the letter of November 1854 were thus reformulated as two 'laws' of electromagnetism. Maxwell justly pointed to the 'recognition of certain mathematical functions as expressing the "electro-tonic state" of Faraday' (*SP*, **1**: 209), as among his claims to originality in 'On Faraday's lines of force'. The introduction of the 'electro-tonic intensity', justified by Faraday's explanation of lines of force and his discussion of electromagnetic induction in 1831, provided a transformation of the two 'theorems' of November 1854 into the formulation of symmetrical 'laws' of electromagnetism in the second part of the published paper. By introducing a new mathematical and physical vocabulary, of 'intensities' and 'quantities', he transformed the meaning of Faraday's theory of lines of force and the electrotonic state. The second part of 'On Faraday's lines of force' is an exercise in conceptual development and transformation.

Maxwell does not appeal to analytical notation in formulating the 'theorems' of the letter of November 1854 or the 'laws' of 'On Faraday's lines of force', but his argument is expressed mathematically. Thus he writes of the 'electro-tonic state', that '[we] find by integration what we may call the *entire electro-tonic intensity round the curve*'; and, with reference to the 'magnetic intensity', he explains that we have 'an integration *round the curve itself* instead of one *over the enclosed surface*' (*SP*, **1**: 206; *LP*, **1**: 374). The mathematical analogy which he draws between the two 'intensities', and the relations between 'magnetic intensity' and 'electric current', and between the 'electro-tonic intensity' and 'magnetic induction', are grounded on relations between curves and surfaces.

In the *Treatise on Electricity and Magnetism* the relations between the electric and magnetic 'intensities' and 'quantities' are expressed in terms of 'Stokes' theorem', an integral theorem which transforms line into surface integrals (see Chapter VII.1). First stated by Thomson in a letter to Stokes in July 1850, the theorem was published by Stokes in his Smith's Prize examination paper in February 1854, where Maxwell, as a candidate for the prize, first encountered it.[21] The form in which Maxwell states these 'theorems' and 'laws' suggests that he had this integral theorem in mind, but his argument is expressed geometrically. The 'laws' are mathematical representations of Faraday's imagery of mutually embracing curves of lines of force and electric currents, expressed as relations between 'intensities' (acting along lines) and 'quantities' (acting through surfaces). In his paper 'On Faraday's lines of force' these field laws rest on geometrical imagery rather than analytic expression by Stokes' theorem.

From the outset, Maxwell had conceived his representation of the magnetic field, based on theorems in potential theory and the geometry of orthogonal surfaces, to be 'geometrical'. In his letter to Thomson in November 1854 he outlines the kernel of the ideas which were to be developed systematically in the second part of 'On Faraday's lines of force': he supposes lines of force to cut 'equipotential surfaces' at right angles so that 'space is now cut up into elements' (*LP*, **1**: 255, 258). As he explains in the introduction to the published paper, '[filling] all space with curves' would provide 'a geometrical model of the physical phenomena'. But he goes on to explain that this geometrical model 'would tell us the *direction* of the force, but we should still require some method of indicating the *intensity* of the force at any point'. To remedy this lacuna, he outlines, in the first part of 'On Faraday's lines of force', the 'physical analogy' of the flow of an incompressible fluid (*SP*, **1**: 155, 158).

In this physical analogy, he supposes 'curves, not as mere lines, but as fine tubes of variable section carrying an incompressible fluid'. In this model, the 'intensity of force as well as its direction' is represented 'by the motion of the fluid in these tubes'. The lines of force form tubes, surfaces directing the motion of an ideal fluid. This physical analogy is explicated without reference to the analytical notation of the mathematical theory of fluids. By expounding 'the purely geometrical idea of the motion of an imaginary fluid', he aims 'to present the mathematical ideas to the mind in an embodied form, as systems of lines or surfaces, and not as mere symbols'.

The espousal of a 'physical analogy' as a 'geometrical idea' served to emphasise one of the main thrusts of the argument of the paper: that his physical geometry of lines of force was not proposed as a 'physical hypothesis'. He aimed to avoid the dangers of 'a premature theory professing to explain the cause of the phenomena'; his argument was a preliminary to 'a mature theory, in which physical facts will be physically explained'. The creation of such a theory was a task for the future. The appeal to a physical analogy enabled him 'to obtain physical ideas without adopting a physical theory'. Moreover, the analogy, as a 'geometrical idea', was an aid to intelligibility: a mathematical formalism would not 'convey the same ideas' as the physical geometry of the 'embodied' system of lines and surfaces (*SP*, **1**: 156, 158–9, 187).

The first part of 'On Faraday's lines of force' is given over to an elaboration of the physical analogy of fluid flow, and its application to some of the phenomena of electricity and magnetism (including paramagnetism and diamagnetism, dielectrics, and the conduction of electricity). But there is no reference to the physical geometry of fluid flow in Maxwell's first account of the genesis of his ideas, his letter to Thomson of November 1854, where he expounds his 'theorems' on lines of force and the magnetic field. His interest in the motion of fluids is first made explicit in his manuscripts and correspondence in April and May 1855, where he sketches out the conditions of stability of the steady motion of an incompressible fluid (*LP*, **1**: 295–9). He set out his results in a letter to Thomson in May 1855 (*LP*, **1**: 309–13); but only in the remark that he was planning to apply Thomson's 'analogy of the steady motion of heat . . . to a more general case to which the laws of heat will not apply' (*LP*, **1**: 307) can there be found a hint that he intended to invoke the motion of an incompressible fluid as a physical analogy for lines of force.

His adoption of the physical analogy of an imaginary incompressible fluid, moving in tubes formed by lines of force, was very likely suggested by Thomson's remark, in his paper 'A mathematical theory of magnetism' (1851), on the 'analogy' between the expression for the 'solenoidal' distribution of magnetism and the 'equation of continuity' in hydrodynamics. Such 'resemblance in mathematical form', Maxwell explains in the introduction to the published paper, was central to his espousal of the method of physical analogy: 'By a physical analogy I mean that partial similarity between the laws of one science and those of another which makes each of them illustrate the other' (*SP*, **1**: 156).

By September 1855, when he next wrote to Thomson, it is apparent that he had it in mind to incorporate the analogy of fluid flow into the argument of his paper.

> The first part of my design is to prove by popular, that is not professedly symbolic, reasoning, the most important propositions about *V* [potential] . . . and to trace the lines of force and surfaces of equal *V*. (*LP*, **1**: 321)

The fluid flow analogy is developed geometrically, in terms of lines, tubes and surfaces, without the symbolic notation used in his drafts on the stability of fluid motion. At around this time he prepared a draft along these lines, setting out propositions on the 'motion of an imponderable and incompressible fluid through a resisting medium' (*LP*, **1**: 337–52). These propositions, reordered, were developed into the argument as it appears in the first two sections of the paper.

By the time he wrote to Thomson in September 1855, the overall scope of the paper, though not its final organisation, was clear to him. This is apparent in one of his most important and revealing declarations about the argument of 'On Faraday's lines of force'; his concise statement of the strategy of the paper, quoted at the beginning of this chapter, merits repetition.

> Now I have been planning and partly executing a system of propositions about lines of force &c which may be *afterwards* applied to Electricity, Heat or Magnetism or Galvanism [Electromagnetism], but which is in itself a collection of purely geometrical truths embodied in geometrical conceptions of lines, surfaces &c. (*LP*, **1**: 320)

By December 1855 he had prepared, for Thomson's perusal and comment, a draft abstract of the first part of the paper (*LP*, **1**: 353–66). Here he gives a clear explanation of his reasoning in introducing fluid flow as an analogy for electricity, rather than simply following Thomson in adopting the analogy of the conduction of heat.

> We should also remember that while the mathematical laws of the conduction of heat derived from the idea of heat as a substance are admitted to be true, the theory of heat has been so modified that we can no longer apply to it the name of substance. It is for this reason that in choosing a concrete form in which to develope the analogy of attraction I have not taken that of heat, as pointed out by Professor Thomson, but at once assumed a purely imaginary fluid as the vehicle of mathematical reasoning.

He wishes to assume, as the basis for geometrical reasoning, a physical analogy having as little associated hypothetical baggage as possible. 'In this way I have endeavoured to make it plain that I am not attempting to establish any physical theory'. The 'method of Physical Analogy', he declares, is a 'method which combines the advantages, while it gets rid of the disadvantages both of premature physical theories and technical mathematical formulæ' (*LP*, **1**: 355–6).

The mode of representation of the magnetic field proposed by Maxwell is justly expressed by the title of his paper. He adopts a theory of the primacy of lines of force in space, a geometrical imagery well exemplified in his account of lines and tubes of force carrying an incompressible fluid. In his draft abstract he notes that

> Faraday treats the distribution of forces in space as the primary phenomenon, and does not insist on any theory as to the nature of the centres of force round which these forces are generally but not always grouped. (*LP*, **1**: 353)

The paper expounds Faraday's concept of the primacy of the field by an exposition of the physical geometry of lines of force, based on an appeal to the flow of an 'imaginary' fluid:

> The substance here treated of . . . is not even a hypothetical fluid which is introduced to explain actual phenomena. It is merely a collection of imaginary properties which may be employed for establishing certain theorems in pure mathematics in a way more intelligible to many minds and more applicable to physical problems than that in which algebraic symbols alone are used. (*SP*, **1**: 160)

True to the ideals of Edinburgh mathematics and philosophy (see Chapter II.3), Maxwell emphasises intelligibility as the aim of mathematics, and geometry as the key to intelligibility. This is an analogy only in the sense of mathematical resemblance; the fluid is not endowed with 'hypothetical' physical properties, which could be translated into properties of lines of force. He aims to avoid any suggestion that the fluid, conceived as a 'collection of imaginary properties', could provide an analogy for a physical theory of lines of force. The model is imaginary and geometrical, an aid to intelligibility. The development of a theory of the physical nature of lines of force, 'in which physical facts will be physically explained' (*SP*, **1**: 159), as demanded by Faraday's own view of the reality of lines of force, was for Maxwell to be a task for the future (see Chapter V.2).

V Models and mechanisms

V.1 Mechanics and molecules: the kinetic theory of gases

Some time between February and May 1859 Maxwell happened to notice, in the *Philosophical Magazine* for February 1859, the translation of a paper by Rudolf Clausius.[1] The paper was Clausius' second contribution to the emergent kinetic theory of gases; and reference in its title to 'the occurrence of molecular motion' may have caught Maxwell's attention. He had only recently completed revising his essay *On the Stability of the Motion of Saturn's Rings* (1859), where he had concluded (see Chapter III.2) that the rings of Saturn consisted in an arrangement of an 'indefinite number of unconnected particles, revolving round the planet with different velocities' (*SP*, **1**: 373). Clausius' paper led him to study the collisions of particles as a means of establishing the properties of gases.

Writing to George Gabriel Stokes in May 1859, he gave a full account of the genesis of his interest in the subject. He emphasises that he had 'taken to the subject for mathematical work', and that he had deduced 'the laws of motion of systems of particles acting on each other only by impact . . . as an exercise in mechanics'. Thus far his investigation can be seen as a continuation of his interest in the collision of particles. But, following Clausius, Maxwell aimed to establish a physical theory of gases. When propositions about the motion of particles were applied to a physical system such as a gas, physical evidence could be invoked, data on gaseous diffusion and viscosity. Thus he hoped 'to be snubbed a little by experiments' (*LP*, **1**: 610–11).

These remarks summarise the strategy of his first paper on the kinetic theory of gases, 'Illustrations of the dynamical theory of gases' (1860). While recognising the distinction between the mechanics of particles and the properties of molecules, the paper applied mathematical propositions on the collisions of elastic spheres to speculations about the physical properties of gas molecules. Beginning as 'an exercise in mechanics', the argument turned to discussion of '"molecular" quantities', derived from the diffusion and viscosity of gases (*LP*, **1**: 610). As he told Stokes,

I have therefore begun at the beginning and drawn up the theory of

> the motions and collisions of particles acting only by impact, applying it to internal friction [viscosity] of gases diffusion of gases and conduction of heat through a gas (without radiation). (*LP*, **1**: 607)

Maxwell places emphasis on the gap between the mathematical theory of the collisions of particles, the dynamics of elastic spheres, and physical speculations about gaseous diffusion and viscosity. As he explained to Stokes the following October, by now preparing his work for publication: 'I intend to arrange my propositions about the molecules of elastic spheres in a manner independent of the speculations about gases'(*LP*, **1**: 619). This remark recalls his comment to Thomson in September 1855 about the strategy of his paper 'On Faraday's lines of force' (see Chapter IV), that his system of geometrical propositions about lines of force could be '*afterwards* applied to Electricity, Heat or Magnetism or Galvanism' (*LP*, **1**: 320).

That he may have intended a loose parallel between the strategies of the two papers is suggested by the introductory remarks to 'Illustrations of the dynamical theory of gases'. The first part of the paper is concerned with 'the motions and collisions of perfectly elastic spheres', and he makes the following programmatic declaration:

> In order to lay the foundations of such investigations on strict mechanical principles, I shall demonstrate the laws of motion of an indefinite number of small, hard, and perfectly elastic spheres acting on one another only during impact.

He then goes on to explain how mechanical propositions about the collisions of elastic spheres could form a 'physical analogy' with the theory of gases.

> If the properties of such a system of bodies are found to correspond to those of gases, an important physical analogy will be established, which may lead to more accurate knowledge of the properties of matter. (*SP*, **1**: 377–8)

The reference to a 'physical analogy' recalls his usage in 'On Faraday's lines of force', but his meaning is different in the two papers.[2] In invoking the 'physical analogy' of elastic spheres, he argues that the properties of colliding elastic spheres will 'correspond to those of gases', and hence establish 'knowledge of the properties of matter'. The role of the physical analogy is to establish an identity between the properties of elastic spheres and those of molecules, and thus to establish the theory of gases. By contrast, his analogy between an incompressible fluid and lines of force is mathematical, and its purpose is to avoid any physical hypothesis. An elastic sphere is a particle with mechanical properties; but Maxwell deliberately eschews physical assumptions in invoking the 'geometrical idea of an imaginary . . . not even hypothetical fluid' (*SP*, **1**: 159–60). Throughout the paper on the kinetic theory of gases, by contrast, he makes reference to the 'hypothesis of elastic particles' (*SP*, **1**: 389), as providing the physical analogy for gas molecules. He presents this as a physical assumption.

> Instead of saying that the particles are hard, spherical, and elastic, we may if we please say that the particles are centres of force, of which the action is insensible except at a certain small distance, when it suddenly appears as a repulsive force of great intensity. It is evident that either assumption will lead to the same results. For the sake of avoiding the repetition of a long phrase about these repulsive forces, I shall proceed upon the assumption of perfectly elastic spherical bodies. (*SP*, **1**: 378)

The two papers have very different strategies.

While Maxwell recognised the importance of separating 'the speculations about gases' from his 'propositions about the motions of elastic spheres' (*LP*, **1**: 619), he aimed to apply his propositions on the collisions of particles to problems of molecular physics, so as to generate a physical theory. From formal propositions on the dynamics of particle collisions, he moves to discuss hypothetical models in molecular physics. That this subject was, in his opinion, truly speculative, can be seen from his remark to Stokes in May 1859: that it was perhaps 'absurd to . . . found arguments upon measurements of strictly '"molecular" quantities before we know whether there be any molecules' (*LP*, **1**: 610).

Despite this caveat, his enthusiasm for engaging in speculations about molecules is not wholly surprising. Clausius, Thomson and Rankine, in developing the science of thermodynamics in the 1850s, had espoused the 'dynamical' or mechanical theory of heat, that heat consists in the motion of the particles of bodies. In his first paper on the kinetic theory of gases, translated in the *Philosophical Magazine* in 1857, Clausius declared his commitment to the mechanical theory of heat. He explained that he had not mentioned the theory in his memoirs on thermodynamics, because he wished to 'separate the conclusions which are deducible from certain general principles from those which presuppose a particular kind of motion'.[3] In expounding his views on thermodynamics, Thomson affirmed this theory of heat, as the title of his paper, 'On the dynamical theory of heat' (1851), made explicit. Rankine published a series of papers proposing a theory of the rotation of 'molecular vortices' in an ether, in an effort to explain the process by which work is converted into heat.[4] The relations between thermodynamics, the mechanical theory of heat and the theory of molecules were explored in the 1850s, much of this work appearing in the *Philosophical Magazine*, which nourished more speculative papers. Hence Maxwell told Stokes that he planned to send his paper 'to the Phil. Magazine which publishes a good deal about the dynamical theories of matter & heat' (*LP*, **1**: 619).

Maxwell describes his theory of gases as 'dynamical', in conformity with contemporary usage. As he explained in his inaugural lecture at King's College London in October 1860, he considered 'Dynamics . . . [to be] the science of the motion of matter as produced by known forces' (*LP*, **1**: 666), a definition he repeated to Stokes in December 1866 in referring to his paper 'On the dynamical theory of gases' (1867): 'Dynamics . . . [is] the theory of the motion of bodies as the result of given forces' (*LP*, **2**: 291). In 1871, following usage suggested in Thomson and Tait's *Treatise on Natural Philosophy* (1867), where the term 'kinetics' is used to refer to 'the science which treats of the action of *force* . . . [which] produces acceleration of relative motion',[5] Maxwell denotes his theory as the 'Kinetic Theory of Gases' (*LP*, **2**: 654).

Having noticed Clausius' paper on the kinetic theory of gases, Maxwell's interest was aroused by its dual focus: particle mechanics and the theory of gases. The study of the physical properties of gases was not unfamiliar to him. His letter to Stokes of May 1859, where he first outlines his work on gas theory, and makes special mention of the viscosity of gases, was written in sequel to a letter of September 1858, inquiring about data on the viscosity of gases and liquids, given by Stokes in a paper on the damping of pendulums (*LP*, **1**: 597–8). Maxwell had been concerned to establish the effect of friction in disturbing the stability of Saturn's rings.

His interest in formulating mechanical theorems on particle collisions very likely arose from his work in revising for publication his Adams Prize essay on Saturn's rings. On reporting to Thomson his conclusion that the ring system was formed of concentric rings of disconnected particles, he remarked that he was unable to calculate the motions of the particles forming the rings: 'the general case of a fortuitous concourse of atoms each having its own orbit & excentricity is a subject above my powers at present' (*LP*, **1**: 555). He amplified this point about computing particle collisions in the published memoir on *Saturn's Rings*: 'When we come to deal with collisions among bodies of unknown number, size, and shape, we can no longer trace the mathematical laws of their motion with any distinctness' (*SP*, **1**: 354). In the memoir on *Saturn's Rings* this remark served as a disclaimer:

> All we can now do is to collect the results of our investigations and to make the best use we can of them in forming an opinion as to the constitution of the actual rings of Saturn which are still in existence and apparently in steady motion, whatever catastrophes may be indicated by the various theories we have attempted. (*SP*, **1**: 354)

These statements suggest that the investigation of the stability of motion of the rings of Saturn alerted him to the problem of calculability. He confronted his inability to compute, 'with any distinctness', the trajectories of the particles constituting the rings of Saturn.

In his paper on the kinetic theory of gases in the *Philosophical Magazine* of February 1859, Clausius calculated the probability of a molecule travelling a given distance without collision, this distance being termed the 'mean length of path'.[6] Maxwell had become interested in probabilities in 1850 (see Chapter VI.2), and this interest no doubt encouraged his ready response to Clausius' paper. In 'Illustrations of the dynamical theory of gases' he advances on Clausius' method by introducing a statistical formula for the distribution of velocities among gas molecules, an expression analogous in form to the distribution function in the theory of errors. As he told Stokes in May 1859

> my particles have not all the same velocity, but the velocities are distributed according to the same formula as the errors are distributed in the theory of 'least squares'. (*LP*, **1**: 610)

The distribution function provides a means of describing the complex motions of colliding particles, enabling him to formulate mechanical theorems expressing the regularity of the motions of molecules. He may have developed his statistical method in response to the problem of calculability he identified in discussing the problem of tracing the motions of the particles in the rings of Saturn. The statistical method does not compute the spatio-temporal trajectories of gas particles, but the distribution function provides a means of describing the pattern of motion of these particles in terms of regularities. The problem of stability of motion of Saturn's rings is transformed into a problem of molecular regularity.[7]

Maxwell himself subsequently drew the link between the two problems. In 1864[8] he attempted to apply his statistical method for the distribution of velocities among particles to the case of the particles forming Saturn's rings. Reviewing his 1859 memoir on *Saturn's Rings* he noted that

> I was then of the opinion that 'When we come to deal with collisions

> among bodies of unknown number size and shape we can no longer trace the mathematical laws of their motion with any distinctness' (§(32)). I propose now to take up the question at this point and to endeavour to throw some light on the theory of a confused assemblage of jostling masses whirling round a large central body . . . . Collisions will occur between these bodies and after collision each body will be projected with a velocity which will carry it into some other part of the cloud of particles, where it will meet with other particles moving with a velocity different from its own. Another collision will thus occur and in this way the jostling of the particles once begun will be carried on throughout the system . . . . The principles by which problems of this kind can be treated were first discussed by Prof[r] Clausius . . . and were applied to several cases in gaseous physics by myself in a paper on the Motions and Collisions of Perfectly Elastic Spheres. (*LP*, **2**: 131–3)

The attempt to apply the statistical method of the theory of gases to the jostling particles in the rings of Saturn proved abortive; but in 1864, if not in 1859, Maxwell perceived a thematic relation between the two problems.

In 'Illustrations of the dynamical theory of gases' Maxwell applies his statistical model of particle collisions to the transport properties of gases, viscosity, diffusion and the conduction of heat. His calculation of the 'mean length of path' between collisions, from data on viscosity and diffusion, forms the centrepiece in his account of the physical properties of gases. He found that the viscosity of a gas was independent of its density, a result he thought surprising, remarking to Stokes that

> This is certainly very unexpected, that the friction should be as great in a rare as in a dense gas. The reason is, that in the rare gas the mean path is greater, so that frictional action extends to greater distances. (*LP*, **1**: 610)

From Stokes' data on viscosity he was able to calculate a value of 1/447 000 of an inch for the mean length of path, and to calculate a value of 1/389 000 of an inch from Thomas Graham's data from experiments on the diffusion of ethylene (*SP*, **1**: 391, 403). He noted that these values were in agreement 'as closely as rough experiments of this kind will permit' (*LP*, **1**: 660).

He obtained another important result in molecular science: the deduction of 'Avogadro's hypothesis' (as it came to be known). As he explained to Stokes in May 1859:

> If two sets of particles act on each other the mean vis viva [kinetic

> energy] of a particle will become the same for both, which implies, that equal volumes of gases at same press. & temp. have the same number of particles, that is, are chemical equivalents. This is one satisfactory result at least. (*LP*, **1**: 610)

It was this result that in 1867 he emphasised to chemists, sceptical of the chemical atomic theory, as an important consequence of the theories 'advocated by physicists from considerations derived from the theory of heat', that is from the kinetic theory of gases (*LP*, **2**: 305).

But one deduction from his theory was more problematic. This he discusses in the third and final part of the paper, concerned with the collision of nonspherical particles. For spherical particles, the motions of rotation are not affected by collisions, but for bodies that are not spherical, there would be a 'relation between the motions of translation and rotation'. He establishes that the kinetic energy of translation would be equal to that of rotation in each system of particles (a result that became known as the equipartition theorem of the equalisation of energy). This had the consequence that there was a conflict between experimental determinations of the specific heats of gases, and their calculation from the kinetic theory:

> by establishing a necessary relation between the motions of translation and rotation of all particles not spherical, we proved that a system of such particles could not possibly satisfy the known relation between the two specific heats of all gases.

Nevertheless, on reviewing the successes of his theory, the calculation of the mean free path and of the conduction of heat in air,[9] and the demonstration of 'Avogadro's hypothesis', he declares that having

> followed the mathematical theory of the collision of hard elastic particles through various cases . . . there seems to be an analogy with the phenomena of gases. (*SP*, **1**: 409)

It was the success of the analogy that was to lead him to continue to pursue the kinetic theory of gases and molecular physics (see Chapters VI.2 and VIII.2). Even though he was, in 1860, to gloomily conclude that the equipartition theorem was fatal to his theory, that this 'result of the dynamical theory, being at variance with experiment, overturns the whole hypothesis, however satisfactory the other results may be' (*LP*, **1**: 660), such pessimism was short-lived. He became more willing to stress the positive conclusions of his molecular theory of gases. The physical analogy of the collisions of elastic particles, commenced as 'an exercise in mechanics', and undertaken 'for mathematical work', became the basis of a theory of molecular physics.

V.2 Ether models: the electromagnetic theory of light

On writing to his friend Cecil James Monro in May 1857, Maxwell remarked that 'I have been grinding at many things and lately during this letter at a Vortical theory of magnetism & electricity which is very crude but has some merits' (*LP*, **1**: 507). This is the first indication of the genesis of his 'theory of molecular vortices' (*SP*, **1**: 451), proposed as a physical representation of lines of force, which is developed in his four-part paper 'On physical lines of force' (1861–2). In this paper he advances from his discussion of the physical geometry of lines of force, in 'On Faraday's lines of force' (1856), to a physical mechanics of the field.

The title of this paper makes an apparent, and probably deliberate, reference to Faraday's paper 'On the physical character of the lines of magnetic force' (1852). There Faraday had affirmed the physical reality of lines of force, and their causal action, as distinct from a purely geometrical treatment of their distribution in space (see Chapter IV.1). This disjunction between geometrical and physical representation is echoed by Maxwell in the titles and strategies of his two papers, 'On Faraday's lines of force' and 'On physical lines of force'. In this latter paper he observes that he had formerly used 'mechanical illustrations to assist the imagination, but not to account for the phenomena' (*SP*, **1**: 452).

The model of lines of flow of an incompressible fluid, deployed in 'On Faraday's lines of force', is a mechanical illustration or analogy (he had termed it a 'physical analogy'), but he had denuded it of physical properties. He had described the analogy as 'not even a hypothetical fluid', but 'a collection of imaginary properties'; its purpose was to assist the imagination, to render the phenomena 'intelligible' (*SP*, **1**: 160). The analogy is illustrative, not explanatory. This appeal to a geometrical model fell short of his aim in writing 'On physical lines of force':

> to examine magnetic phenomena from a mechanical point of view, and to determine what tensions in, or motions of, a medium are capable of producing the mechanical phenomena observed.
>
> (*SP*, **1**: 452)

In 'On Faraday's lines of force' he had made reference to Thomson's 1847 representation of electric and magnetic forces by strains in an elastic solid, in looking to a mechanical, explanatory theory of Faraday's concept of the electro-tonic state:

> By a careful study of the laws of elastic solids and of the motions of

> viscous fluids, I hope to discover a method of forming a mechanical conception of this electro-tonic state adapted to general reasoning.
> (*SP*, **1**: 188)

But Thomson had merely provided an 'allegory about incompressible elastic solids', as Maxwell had observed in 1855 (*LP*, **1**: 320), a point he amplifies in 'On physical lines of force': '[Thomson] does not attempt to explain the origin of the observed forces by the effects due to these strains in the elastic solid, but makes use of the mathematical analogies . . . to assist the imagination' (*SP*, **1**: 453). Thomson's analogy is illustrative, not explanatory, having a similar role to Maxwell's geometrical model of an imaginary, incompressible fluid.

Maxwell was now seeking a physical analogy in the sense of a causal, explanatory model, as in 'Illustrations of the dynamical theory of gases'. He aimed to provide an explanatory, rather than an illustrative, mechanical analogy for lines of force. He found this analogy in Thomson's explanation of the Faraday magneto-optic rotation: 'Thomson has pointed out that the cause of the magnetic action on light must be a real rotation going on in the magnetic field' (*SP*, **1**: 505), the rotation of light being produced by the rotation of molecular vortices in an ether. In the 'theory of molecular vortices' proposed in 'On physical lines of force', the forces produced by the centrifugal rotation of the vortices are isomorphic to magnetic forces. The mechanical model of molecular vortices can therefore be identified with the magnetic field, and provides an explanatory physical analogy.[10]

Thomson's explanation of the magneto-optic effect was proposed in 1856, his paper being reprinted in the *Philosophical Magazine* in March 1857, when it was soon noticed by Maxwell. Thomson had argued that the Faraday rotation was to be explained on the assumption that magnetic lines of force were axes of rotation in an ether, and that this medium contained particles set in circular motion by the magnetic field. The vibrations constituting light are supposed to interact with circular motions of the elements of the medium, which have axes parallel to the magnetic field. On being traversed by a polarised light ray, the rotation of the vortices would produce the Faraday effect. Thomson aimed to apply the theory to all the phenomena of electromagnetism. He found the model for his 'dynamical illustration' of the electromagnetic field in the dynamical theory of heat, making reference to Rankine's hypothesis of the rotary motion of 'molecular vortices' as the cause of heat.

> The explanation of all phenomena of electro-magnetic attraction or repulsion, and of electro-magnetic induction, is to be looked for simply

> in the inertia and pressure of the matter of which the motions constitute heat.[11]

Maxwell's interest in vortices, first declared in his letter to Monro in May 1857, was clearly sparked by Thomson's paper. Writing to Faraday in November 1857, he looked to the extension of the theory of the field towards a 'possible confirmation of the physical nature of magnetic lines of force'. Thomson's explanation of the Faraday effect by a vortical motion in a medium offered hope of achieving this objective.

> But there are questions relating to the connexion between magneto electricity and certain mechanical effects which seem to me opening up quite a new road to the establishment of principles in electricity and a possible confirmation of the physical nature of magnetic lines of force. Professor W. Thomson seems to have some new lights on this subject.
> (*LP*, **1**: 552)

These were the ideas that engaged Maxwell's attention as early as May 1857, leading him to his 'theory of molecular vortices' in 'On physical lines of force', which provides a physical mechanics of an ambient medium as the substratum of the electromagnetic field. Writing to Thomson in December 1861, he affirmed that 'I have been trying to develope the dynamical theory of magnetism as an affection of the whole magnetic field according to the views stated by you', citing Thomson's paper on the Faraday magneto-optic effect (*LP*, **1**: 692).

During 1857 Thomson became preoccupied with speculations on hydrodynamics and vortices. In a series of letters to Stokes he pondered on the mechanism of the rotation of particles within a medium: 'What I am most anxious to make out however is the mutual action of motes, separated by a perfect liquid'. He recollected a paper of Maxwell's, a 'curious illustration' (as he describes it)[12] of rotary motion: the generation of rotation, round a horizontal axis, in a slip of paper falling through the air, causing it to follow a curved trajectory (*LP*, **1**: 213–18; *SP*, **1**: 115–18). Maxwell was drawn into the discussion, and in a letter to Thomson in November 1857 he summarises his explanation of the trajectory of the slip of paper, that its rotary motion was caused by air resistance. He goes on to discuss the generation of eddies in a fluid, an issue raised by Thomson:

> Now I do not see why the unstable motion of a perfect fluid should not produce eddies which . . . remain in the fluid in a state of subdivision which is as nearly that of molecular vortices as any finite motion can be.
> (*LP*, **1**: 560–2)

Molecular vortices were therefore dynamically conceivable. His continued enthusiasm for the theory can be seen in his discussion of magnetism in a letter to Thomson in January 1858. Here he supposes 'magnetism to consist in the revolution or rotation of any material thing', and he proceeds to outline the design of an experiment on a freely rotating magnet, in an attempt to establish the effect of revolving vortices within the magnet (*LP*, **1**: 579–80). In 'On physical lines of force' he mentions undertaking such experiments (*SP*, **1**: 485–6n); and in the *Treatise on Electricity and Magnetism* (1873) he describes the apparatus he had used, 'which I had constructed in 1861', and which is still extant.[13] The apparatus is designed to detect molecular vortices by determining the effect of the angular momentum of revolving vortices on the free rotation of an electromagnet. An electromagnet can rotate about a horizontal axis within an armature, which revolves about a vertical axis. If the current in the coil carries momentum, then the coil would precess about the vertical axis. However, he found 'no evidence of any change' (*Treatise*, **2**: 202–4 (§575)). Writing to Thomson in December 1861 he had reported that

> I find that unless the diameter of the vortices is sensible, no result is likely to be obtained by making a magnet revolve freely about an axis perp. to the magnetic axis. (*LP*, **1**: 698)

The effect would be unobservably small.[14]

Maxwell's 'theory of molecular vortices', published in the first two parts of 'On physical lines of force' in the March, April and May 1861 numbers of the *Philosophical Magazine*, was of long gestation. But there is no mention of the theory in his correspondence from January 1858 until October 1861, after its publication. It seems likely that he put this work aside. His proposal of a physical theory of lines of force, based on speculations about 'molecular vortices' in a fluid medium, may have been encouraged by the success of his 'physical analogy' of particles in collision, in generating '"molecular" quantities' in the theory of gases (*LP*, **1**: 610). As with 'Illustrations of the dynamical theory of gases', he chose the *Philosophical Magazine* as the appropriate venue for publication of a paper advancing a speculative physical theory.

'On physical lines of force' sets out a mechanical representation of Faraday's theory of the field, explicated in terms of 'the theory of molecular vortices'. Postulating a physical mechanics of a medium as the substratum of the electromagnetic field, this theory embodies Maxwell's first attempt to realise Faraday's programme, as he construes it: to find 'the seat of the phenomena in real actions going on in the medium' (*Treatise* **1**: x).

Writing to Faraday in October 1861, after the publication of the first two parts of the paper, he sought to place his theory firmly within the Faradayan paradigm.

> When I began to study electricity mathematically, I avoided all the old traditions about forces acting at a distance, and after reading your papers as a first step to right thinking, I read the others, interpreting as I went on, but never allowing myself to explain anything by these forces. It is because I put off reading about electricity till I could do it without prejudice, that I think I have been able to get hold of some of your ideas such as the electrotonic state, action of contiguous parts &c and my chief object in writing to you is to ascertain if I have got the same ideas which led you to see your way into things or whether I have no right to call my notions by your names. (*LP*, **1**: 688)

This statement, and his declaration to Thomson in December 1861, that he had been striving to develop 'the dynamical theory of magnetism . . . according to the views stated by you' (*LP*, **1**: 692), provide emphatic indications of the antecedents and intentions of Maxwell's paper.

The first part of the paper is concerned with the application of the theory of molecular vortices to magnetism. He investigates the effect of vortices in a fluid, arranged in tubes or filaments so as to correspond to the lines of force (the magnetic field), and rotating so as to generate magnetic forces. The angular velocity of the vortices corresponds to the magnetic field intensity. The rotational motion generates centrifugal forces in the vortex filaments, which expand equatorially and contract along their lengths, generating a stress in the fluid medium filled by the vortex filaments (*SP*, **1**: 454–5).

In the second part of the paper, he notes that he had thus far shown that 'the hypothesis of vortices . . . gave a probable answer' to the question of the 'mechanical cause' of the required stresses in the medium. But to explain electromagnetism and electromagnetic induction required an extension in the scope of the theory: 'We have, in fact, now come to inquire into the physical connexion of these vortices with electric currents'. This leads him to an expansion of the model of molecular vortices as initially postulated. He states the problem in mechanical terms: to explain the coupling of the vortices. He proposes a mechanical analogy to explain the rotation of neighbouring vortices revolving in the same direction about parallel axes. By analogy to a 'mechanism' in which an 'idle wheel' is placed between two wheels that are intended to revolve in the same direction, he suggests that

> a layer of particles, acting as idle wheels, is interposed between each vortex and the next, so that each vortex has a tendency to make the neighbouring vortices revolve in the same direction with itself.
>
> (*SP*, **1**: 467–8)

The figure he uses to provide a physical representation of this model of rotating vortices and idle wheel particles (Fig. V.1), depicts the 'magneto-electric medium' in cross-section as an array of regular hexagons, like a honeycomb. This figure suggests that he envisaged a three-dimensional cellular structure, the vortex cells being (irregular) rhombic dodecahedra (resulting from the densest packing of equal spheres). Each molecular vortex is surrounded by a layer of 'idle wheel' particles; these particles would be subjected to translational motion if adjacent rotating vortices had different angular velocities.

The mechanical model is proposed as an electromagnetic analogy, to explain electromagnetic induction and the generation of electric currents. Changes in the rotational motion of the vortices would be produced by the tangential forces exerted by the idle wheel particles on the surface of the vortices; and there would be inertial reaction forces exerted by the vortices on the particles. The electromotive force 'arises from the action between the vortices and the interposed particles, when the velocity of rotation is altered in any part of the field'. He explains that the 'relation between the alterations of motion . . . and the forces exerted on the layers of particles between the vortices', represents 'in the language of our hypothesis, the relation between changes in the state of the magnetic field and the electromotive forces thereby brought into play'. The translational motion of the idle wheel particles corresponds to the flow of an electric current in an inhomogeneous magnetic field: 'an electric current is represented by the transference of the moveable particles interposed between the neighbouring vortices'. The mechanical model thus has electromagnetic correlates (*SP*, **1**: 471, 475–6, 478).[15]

The electro-tonic state has an important role in the field equations which he deduces by invoking the relation between the mechanical model and its electromagnetic correlates. The interpretation of the electro-tonic state, originally suggested by Faraday to explain electromagnetic induction (see Chapter IV.1), is embedded in the mechanical model which provides the basis for the conceptualisation. The electro-tonic state is 'what the electromotive force would be if the currents . . . had started instantaneously from rest'. He defines it as an '*impulse*', and he terms it the '*reduced momentum*' of the vortices, an

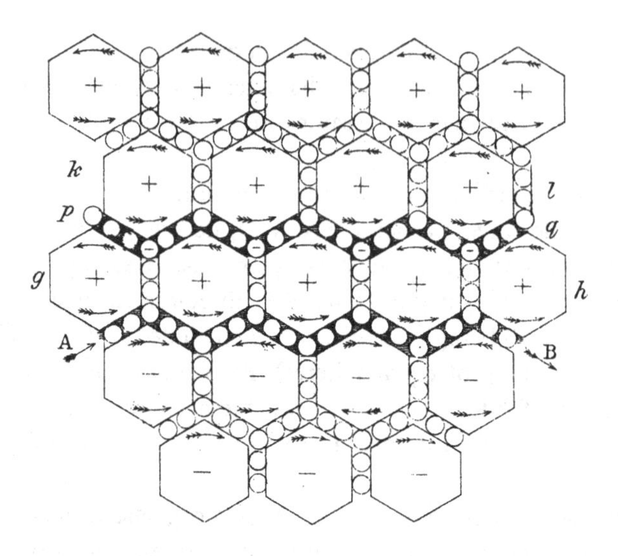

Fig. V.1. Maxwell's ether model, described in 'On physical lines of force' (1861–2), of vortices (whose rotation represents the magnetic field), separated by 'idle wheel' particles (representing electricity), which enable adjacent rows of vortices to rotate about parallel axes in the same direction. The electric current flows from *A* to *B*, and the row of vortices *gh* is set in motion in an anti-clockwise direction (+), and acts on the layer of particles *pq*. These will rotate in a clockwise direction (−), and will move from right to left, in the opposite direction to the primary current, and so form an induced electric current. If this current is checked by the electrical resistance of the medium, the rotating particles *pq* will act on the next row of vortices *kl*, which will rotate in an anti-clockwise direction (+). If the primary current *AB* is stopped, the vortices *gh* will cease to rotate, but the momentum of the vortices *kl* will tend to move the layer of particles *pq* from left to right, in the direction of the primary current. If this motion is resisted by the medium, the motion of the vortices beyond *pq* will cease. The figure contains a drafting error: the arrows on the vortices below *AB* should be marked to represent rotation in a clockwise direction (−). From *SP*, **1**: Pl. VIII facing p. 488.

impulse generated when the vortices are brought to rest by the cessation of flow of the idle wheel particles. This provides an interpretation, 'from a mechanical point of view', of the relation between the electro-tonic state and electromotive force (*SP*, **1**: 478–9).

Maxwell concludes the second part of his paper by reflecting on the status of his mechanical model of vortices and idle wheel particles. While he considered his basic 'hypothesis of vortices' to be 'probable' (*SP*, **1**: 468), he distinguishes the theory of molecular vortices itself from the ether model he had developed in further application of the theory. He describes the ether model of idle wheel particles and vortices, advanced in the second part of the paper, as a 'provisional and temporary' hypothesis. He considers the rolling contact model to be 'awkward'; and he states unequivocally that 'I do not bring it forward as a mode of connexion existing in nature, or even as that which I would willingly assent to as an electrical hypothesis'. On the positive side, he maintains that the model is 'mechanically conceivable', and its application demonstrates the possibility of a mechanical explanation of the electromagnetic field: 'it serves to bring out the actual mechanical connexions between the known electro-magnetic phenomena' (*SP*, **1**: 486).

These reflections on the tentative and illustrative status of the ether model were very likely intended as a conclusion to the paper as originally envisaged. The extension to electrostatics, documented in the third part of the paper published in January 1862, was yet to be accomplished. His comments on the rationale of the ether model may be compared with his concluding remarks on the theory of molecules as elastic spheres in 'Illustrations of the dynamical theory of gases'. There the problem was the conflict between the mathematical theory and experimental data of specific heats; here the problem was the justification of the elaborate edifice on which the theoretical argument rests.

In setting out his objectives in 'On physical lines of force', he makes no reference to the experimental discovery (the Faraday rotation) and its explanation (by Thomson), which had led him to formulate his theory. The paper is not presented as a response to the fruitfulness of Thomson's speculation. The vortices are initially envisaged as filaments in a fluid, and subsequently as vortex cells, to explain magnetism and electric currents. The property of elasticity, associated with an optical ether, is only introduced in the third part of the paper. 'The undulatory theory of light requires us to admit this kind of elasticity in the luminiferous medium, in order to account for transverse vibrations . . . the magneto-electric medium possesses the same property'. Having suggested that the vortices 'consist of the same matter the vibrations of which constitute light' (*SP*, **1**: 489, 506), he turns, in the fourth part of the paper, to consider the magneto-optic rotation, and finally cites Thomson.

Maxwell offers the paper in fulfilment of his own scientific programme. Having achieved a physical geometry of lines of force, he now proposes to advance to a physical mechanics, discussing Faraday's theory 'from a mechanical point of view', to 'clear the way for speculation . . . by investigating . . . the action of a medium'. The argument is justified as an elaboration of a theoretical perspective, the proposal of 'a theory which, if not true, can only be proved to be erroneous by experiments which will greatly enlarge our knowledge of this part of physics' (*SP*, **1**: 452). In the introduction to 'On Faraday's lines of force' he confesses that he had 'hardly made a single experiment' in the science; his argument is mathematical and did not rest on new experimental results (*SP*, **1**: 157). The first two parts of 'On physical lines of force' also offer a theoretical perspective, and he echoes the language of the earlier paper: 'We have now shown in what way electro-magnetic phenomena may be imitated by an imaginary system of molecular vortices' (*SP*, **1**: 488).

Maxwell had yet to attempt to extend the theory of molecular vortices to electrostatics; this he accomplished in the third part of the paper. Two problems which stood in the way of the success of the vortex theory were then resolved: the explanation of static electric charge, and the representation of the mechanism of the interaction of the idle wheel particles with the fluid surfaces of the vortices.[16] He suggests that the cells of the vortex medium were endowed with elastic properties and interact elastically with the idle wheel particles. He supposes the 'magneto-electric medium' to be divided into spherical 'cells separated by partitions, formed of a stratum of particles which play the part of electricity'. The motion of these electric particles will distort the cells and give rise to an equal and opposite force originating from the elasticity of the cells. The assignment of elastic properties to the vortex cells explains how the idle wheel particles could exert tangential forces upon the solid surfaces of the spherical cells.

The mechanical structure of the elastic ether has an electrical analogue. The elastic distortion of the cells has as its analogue a 'displacement' of electricity in the cell, which represents the 'polarization' of the particles of the 'magneto-electric medium'. The effect of this action on the dielectric medium is to produce a 'general displacement of the electricity in a certain direction', thus explaining the accumulation of static electrical charge (*SP*, **1**: 489–92).

Writing to Faraday in October 1861, to explain the development of his theory, and to 'ascertain if I have got the same ideas which led you to see your way into things' (*LP*, **1**: 688), Maxwell gives an account of his theory of electrostatics.

> My theory of electrical forces is that they are called into play in insulating media by *slight* electric displacements, which put certain small portions of the medium into a state of distortion which, being resisted by the *elasticity* of the medium, produces an electromotive force. A spherical cell would by such a displacement be distorted thus where the
>
> 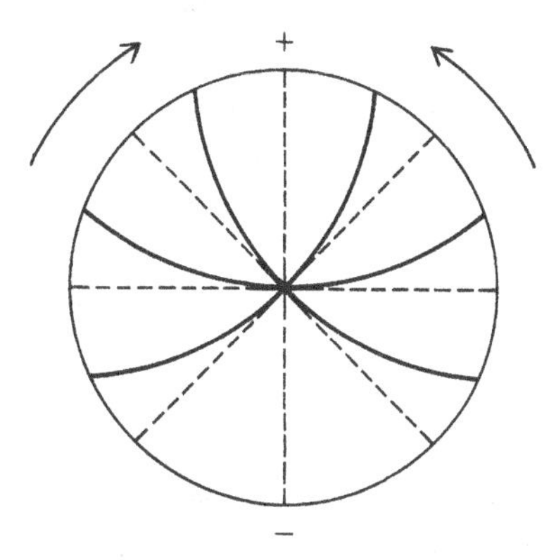
>
>
> curved lines represent diameters originally straight, but now curved.
>
> I suppose the elasticity of the sphere to react on the electrical matter surrounding it, and press it downwards. (*LP*, **1**: 684–5)

Adopted as a representation of Faraday's concept of the polarisation of the dielectric medium, 'a state of polarization of its parts', as Maxwell describes it, the concept of 'electric displacement' is sanctioned by the appeal to the cellular ether model. Its introduction provides a mechanical illustration of Faraday's theory of electrostatic action between the contiguous particles of the dielectric medium, in which the particles of the dielectric were supposed to be in a 'polarized state', 'assuming positive and negative points or parts'.[17] As Maxwell describes the concept in the third part of 'On physical lines of force':

> Electromotive force acting on a dielectric produces a state of polarization of its parts similar in distribution to the polarity of the particles of iron under the influence of a magnet, and, like the magnetic polarization, capable of being described as a state in which every particle has its poles in opposite conditions . . . . The effect of this action on the whole dielectric mass is to produce a general displacement of the electricity in a certain direction. (*SP*, **1**: 491)

The incorporation of electric displacement into the field equations arose from Maxwell's development of the Faradayan concept of dielectric polarisation. This was, moreover, a physical theory of the field because the field of force is represented mechanically by a theory of the transmission of force between the particles of the elastic vortex medium. As he informed Faraday, he had 'endeavoured to form a mechanical conception of the part played by the particles of air, glass or other dielectric in the electric field' (*LP*, **1**: 683). The displacement current was introduced into the 'theory of molecular vortices' to allow for the elasticity of the vortex medium, thus providing an explanation of electric charge. The elasticity of the vortex medium provides the mechanical foundation for the introduction of the displacement current.[18] This is, therefore, a physical mechanics of the ether, grounded on Faraday's 'theory of action exerted between contiguous parts of a medium' (*Treatise*, **1**: 99 (§95)).

The theory had an unexpected consequence. The supposition of the elastic property of the 'magneto-electric medium' allowed for the propagation of elastic shear waves within the medium. From the model, Maxwell established that the velocity of propagation of transverse waves in the medium was equal to the ratio of electrical units (see Chapter III.3). The numerical value of the ratio of units had been established by Weber and Kohlrausch, and Maxwell found agreement between their value and the velocity of light. When in October 1861 he wrote to inform Faraday of his conclusion that 'the luminiferous and electromagnetic medium are one', this claim was based on his theoretical argument, supported by the numerical value of the ratio of electrical units. 'I worked out the formulæ in the country, before seeing Webers number', he told Faraday, implying that he had been unaware that the value of the ratio of units, and hence the velocity of propagation of waves in the medium, was close to the measured velocity of light. He concluded that: 'This coincidence is not merely numerical' (*LP*, **1**: 685–6).

He repeats this claim to Thomson in December 1861: 'I made out the equations in the country before I had any suspicion of the nearness between the two values of the propagation of magnetic effects and that of light' (*LP*, **1**: 695). In the published paper, he drew the conclusion in the most emphatic terms:

> The velocity of transverse undulations in our hypothetical medium . . . agrees so exactly with the velocity of light . . . that we can scarcely avoid the inference that *light consists in the transverse undulations of the same medium which is the cause of electric and magnetic phenomena.*
>
> (*SP*, **1**: 500)

2 The electromagnetic theory of light

In 'On physical lines of force' Maxwell does not conceive light waves as electromagnetic waves. Though he does so subsequently (see Chapter VI.1), he still conceives the generation of light to be a mechanical rather than an electromagnetic process. Optical and electromagnetic processes are conceptualised in mechanical terms: the relation between optics and electromagnetism is expressed by the molecular connection between ether and matter, in terms of different motions in the ether. Moreover, in his first 'electromechanical'[19] theory of light, the elastic properties of the mechanical medium establish the connection between optical and electromagnetic phenomena. Maxwell may have adjusted the parameters of his model in obtaining the result that the velocity of waves in the magneto-electric medium was equal to the ratio of electrical units; but the numerical agreement between the measured value of the ratio of units and the observed velocity of light was unexpected, and was between values established experimentally.[20]

Maxwell was able to derive two consequences from the theory, which further supported the claimed connection between electromagnetism and optics. The first was a relation between the dielectric constant and the refractive index of a medium. As he had informed Faraday in October 1861:

> Supposing the luminous and the electromagnetic phenomena to be similarly modified by the presence of gross matter, my theory says that the inductive capacity (static) is equal to the square of the index of refraction, divided by the coefficient of magnetic induction (air = 1).
>
> (*LP*, **1**: 686–7)

He subsequently asked Thomson for information about 'any good measures of dielectric capacity of transparent substances' (*LP*, **1**: 696). The experimental support was approximate (*SP*, **1**: 500–1, 582–3), and he planned his own determination of dielectric constants (*LP*, **1**: 696, 702, 710). In 1870–1 he undertook experiments in an attempt to establish the value of the refractive index of paraffin (*LP*, **2**: 543, 627). Only in the 1870s did Ludwig Boltzmann provide strong confirmatory evidence to support the result.[21]

But the second consequence of his electromagnetic theory of light bore more immediate fruit. To complete his theory of physical lines of force, he wished to give a quantitative treatment of the Faraday effect in terms of the theory of molecular vortices. Thomson's suggestion that the magneto-optic rotation could be explained by the rotation of vortices had initiated his investigation, and could now be discussed in terms of the newly created electromagnetic theory of light. This formed the argument of the fourth and final part of 'On physical lines of force', published in the *Philosophical Magazine* in February 1862.

> We have now to investigate whether the hypothesis developed in this paper – that magnetic force is due to the centrifugal force of small vortices, and that these vortices consist of the same matter the vibrations of which constitute light – leads to any conclusions as to the effect of magnetism on polarized light. (*SP*, **1**: 506)

He had first raised the issue with Faraday in October 1861.

> I have also examined the theory of the passage of light through a medium filled with magnetic vortices, and find that the rotation of the plane of polarization is in the same direction with that of the vortices, that it varies inversely as the *square* of the wavelength (as is shown by experiment) and that its amount is proportional to the *diameter* of the vortices. (*LP*, **1**: 687)

But his account of the Faraday effect was incomplete:

> I have not yet found any determination of the rotation of the plane of polarization by magnetism in which the absolute intensity of magnetism at the place of the transparent body was given. I hope to find such a statement by searching in libraries, but perhaps you may be able to put me on the right track. (*LP*, **1**: 684)

Faraday annotated the letter with the name 'Verdet'; and it is possible that he drew Maxwell's attention to the experiments of Émile Verdet on the magneto-optic effect. Certainly by the following December, when he wrote to Thomson, Maxwell's account of the Faraday effect had broadened to encompass Verdet's work.

> I have also examined the propagation of light through a medium containing vortices, and I find that the *only* effect is the rotation of the plane of polarization in the *same* direction as the angular momentum of all the vortices the rotation being proportional to
>
> A the thickness of the medium
> B the magnetic intensity along the axis
> C the index of refraction in the medium
> D inversely as the square of the wave length in air
> F directly as the radius of the vortices
> G . . . . .the magnetic capacity.

Of these, he notes, 'A & B are proved by Verdet' (*LP*, **1**: 697–8).[22] With the alphabetisation corrected, this is essentially how the argument is presented in the paper. In considering the magneto-optic effect in the terms of the theory of molecular vortices, he is able to appeal to experimental results, notably those provided by Verdet, in support of his theory (*SP*, **1**: 506–13).

Reviewing the theory of molecular vortices in this final part of his paper, Maxwell emphasises a key feature of his ether model: it enabled him to explain the relation between 'magnetism as a phenomenon of rotation' (as implied by the Faraday effect) and 'electric currents as consisting of the actual translation of particles'. He argues that electric currents must be conceived as linear in character (he gives electrolysis as an example), while he maintains that magnetism is a rotational phenomenon as is shown by 'the rotation of the plane of polarized light when transmitted along the lines of magnetic force'. Assuming that 'the direct effects of a rotatory cause must be themselves rotatory', the Faraday effect indicated that there was 'a real rotation going on in the magnetic field' (*SP*, **1**: 502–5).[23]

This argument seemed to Maxwell to provide compelling evidence in favour of the basic physical assumption of 'On physical lines of force': the 'theory of molecular vortices'. As his emphatic statement in the *Treatise* makes clear, he did not waver from the view that the Faraday effect demonstrated that 'some phenomenon of rotation is going on in the magnetic field'. This rotation implied that there was 'some kind of mechanism' in the magnetic field, a mechanism performed by 'a great number of very small portions of matter, each rotating on its own axis'. While the Faraday effect implied rotation and a mechanism, and rotating vortices, it did not specify any particular mechanical model. As he continues to affirm in the *Treatise* (see Chapter VI.1), the 'idle wheel' ether model, his 'working model of this mechanism', was merely a 'clumsy' representation of a conceivable mechanism; but, though awkward, it had the virtue of demonstrating the possibility of constructing a mechanical model (*Treatise*, **2**: 416–17 (§831)).

Proposed as a physical theory of lines of force, the model of molecular vortices evolved as the paper progressed: from filaments in a fluid, to vortex cells, to elastic cells. With the extraordinary innovation of the electromagnetic theory of light, one of the most striking and significant results in the history of physics, the hypothesis of vortices had proved unexpectedly productive. With its further application to subsume the Faraday magneto-optic rotation, the theory developed in 'On physical lines of force' had ultimately proved to be more productive and complete than the 'physical analogy' of colliding particles of 'Illustrations of the dynamical theory of gases'. Both papers were fuelled by the same impulse: to advance beyond a 'geometrical model', or an 'exercise in mechanics', to provide a physical theory in the form of a mechanical mode of representation, whether of an ether model or the dynamics of colliding particles. Both papers had limitations, perceived and indeed highlighted by Maxwell at the time. In the case of the model of colliding elastic particles in 'Illustrations of the dynamical theory of gases', the most obvious and immediate difficulty arose from the consequences of the equipartition theorem; but other problems subsequently became apparent (see Chapters VI.2 and VIII.2). The argument of 'On physical lines of force' was limited by a philosophical problem, an issue of scientific explanation: the general validity of results tied so closely to a mechanical model of the ether (see Chapter VI.1).

VI Dynamical and statistical explanation

VI.1 The dynamical theory of the electromagnetic field

In December 1861, even before the two final parts of 'On physical lines of force' had been published, Maxwell wrote to a Cambridge friend informing him that 'I am trying to form an exact mathematical expression for all that is known about electro-magnetism without the aid of hypothesis' (*LP*, **1**: 703). In the paper he had considered the 'hypothesis of vortices' to be 'probable', but the ether model of vortex cells and idle wheel particles to be 'awkward', a 'provisional and temporary hypothesis' (*SP*, **1**: 468, 486). His caution about the rationale of his 'theory of molecular vortices' was made explicit, and led him to seek to transcend its limitations.

In subsequent comment he mixed lofty disengagement from the 'awkward' and 'provisional' assumptions of the ether model, with a defence of the validity of the 'hypothesis of vortices', asserting the legitimacy of the theory as an explanatory strategy. Writing to William Thomson in October 1864, he was on the defensive in response to what he took to be Thomson's offhand dismissal of the theory. 'The tendency in my rotatory theory of magnetism was towards the to me inconceivable and ∴ no doubt to the misty though why you put a *c* after the *y* I cannot see'.[1] But by this time he was about to submit his paper 'A dynamical theory of the electromagnetic field' (1865) to the Royal Society, and he left Thomson in no doubt that he had been able to discard, and transcend, the 'inconceivable' and 'misty' ether model.

> I can find the velocity of transmission of electromagnetic disturbances indep^t of any hypothesis now & it is = *v* [the ratio of electrical units] and the disturbances must be transverse to the direction of propagation or there is no propagation thereof. (*LP*, **2**: 180)

He elaborated this point in a letter written the same day to Stokes, as Secretary of the Royal Society.

> I have now got materials for calculating the velocity of transmission of a magnetic disturbance through air founded on experimental evidence without any hypothesis about the structure of the medium or any mechanical explanation of electricity or magnetism.

> The result is that only transverse disturbances can be propagated and that the velocity is that found by Weber and Kohlrausch which is nearly that of light. (*LP*, **2**: 187–8)

He asserts that he has now established, independently of 'any hypothesis', that electromagnetic disturbances are propagated as transverse waves, like light; that the velocity of propagation of electromagnetic waves is equal to the ratio of electrical units; and that this ratio had been measured by Weber and Kohlrausch, and had a value nearly that of the velocity of light (see Chapter V.2).

The dimensional notation proposed in Maxwell's paper 'On the elementary relations of electrical quantities', written in collaboration with Fleeming Jenkin, and which was included in the 1863 report of the Committee on electrical standards, may have helped to foster the strategy of 'A dynamical theory of the electromagnetic field' (1865). The dimensional notation established a phenomenological link between v, the ratio of electrostatic and electromagnetic units, and the velocity of light. In the autumn and winter of 1864–5 Maxwell corresponded with Thomson over methods of obtaining v, finally settling on equilibrating the electrostatic attraction of two discs with the electromagnetic repulsion of two coils (see Chapter III.3). Writing in September 1864 to Charles Hockin, an assistant in the work on standards and with whose collaboration he undertook the experiments, he associates the measurement of v with the successful conclusion of his new paper. He states that he had 'cleared the electromagnetic theory of light from all unwarrantable assumption, so we may safely determine the velocity of light by measuring the attraction between bodies kept at a given difference of potential'(*LP*, **2**: 164).

Maxwell thus claims to have obtained his 'Electromagnetic Theory of Light', as he now terms it (*LP*, **2**: 194; *SP*, **1**: 577), without reference to the ether model or to the mechanical analogies employed in 'On physical lines of force'. He makes his most emphatic statement of the postulates of his newly wrought theory in the abstract of his paper, published in the *Proceedings of the Royal Society*.

> What, then, is light according to the electromagnetic theory? It consists of alternate and opposite rapidly recurring transverse magnetic disturbances, accompanied with electric displacements, the direction of the electric displacement being at right angles to the magnetic disturbance, and both at right angles to the direction of the ray. (*LP*, **2**: 195)

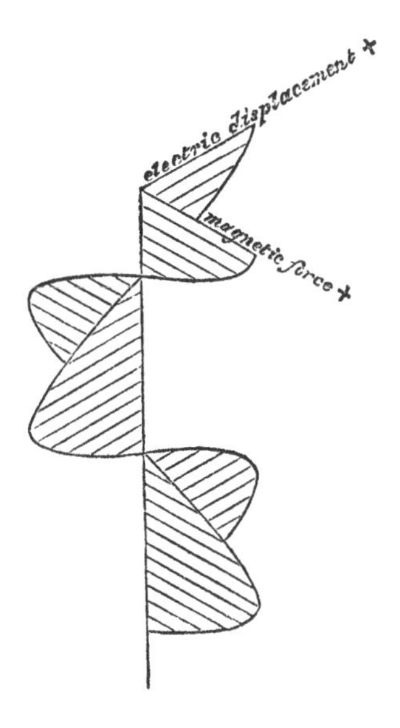

Fig. VI.1. The electromagnetic theory of a light ray. From *Treatise*, **2**: 390 (§791). The figure illustrates Maxwell's theory that 'light is an electromagnetic disturbance, propagated in the same medium through which other electromagnetic actions are transmitted' (*Treatise*, **2**: 387 (§786)).

In the *Treatise on Electricity and Magnetism* (1873) he gives a visual representation of this electromagnetic theory of a light ray (Fig. VI.1). But while he dispensed with a mechanical model of the ether in deriving his 'Electromagnetic Theory of Light', he does not discard appeal to the dynamics of an ethereal medium. The claim that 'light and magnetism are affections of the same substance' (*SP*, **1**: 580) was a cardinal feature of his theory.

Maxwell begins his paper 'A dynamical theory of the electromagnetic field' by giving a definitive statement of the scope and nature of his argument.

> The theory I propose may therefore be called a theory of the *Electromagnetic Field*, because it has to do with the space in the neighbourhood of the electric or magnetic bodies, and it may be called a *Dynamical* Theory, because it assumes that in that space there is matter in motion, by which the observed electromagnetic phenomena are produced.
>
> (*SP*, **1**: 527)

The theory is based on the propagation of vibrations in the 'aethereal medium filling space and permeating bodies'. The transmission of motion in the ether required that its parts be set in motion, and that its 'connexions must be capable of a certain kind of elastic yielding'. Following the establishment of energy physics in the 1850s, Maxwell expounds his dynamical theory in terms of energy transformations in the ether. Using terms introduced by Rankine in 1853,[2] he distinguishes between 'the "actual" [kinetic] energy depending on the motion of its parts, and "potential" energy, consisting of the work which the medium will do in recovering from displacement in virtue of its elasticity'. The theory of energy transformations in the medium provides the model for the electromagnetic theory of light: 'The propagation of undulations consists in the continual transformation of one of these forms of energy into the other alternately' (*SP*, **1**: 528–9).

Electromagnetic phenomena are considered as effects of the dynamics of the ethereal medium. He once again appeals to the Faraday magneto-optic effect as providing evidence that there may be a 'motion of the ethereal medium going on wherever magnetic effects are observed'; and notes that 'this motion is one of rotation, having the direction of the magnetic force as its axis'. He explains electromotive force as 'the force called into play during the communication of motion from one part of the medium to another'; and when electromotive force acts on a dielectric 'it produces a state of polarization of its parts . . . a general displacement of electricity in a certain direction . . . [which is] a kind of elastic yielding to the action of the force' (*SP*, **1**: 529–31). He therefore concludes that

> we are led to the conception of a complicated mechanism capable of a vast variety of motion, but at the same time so connected that the motion of one part depends . . . on the motion of other parts, these motions being communicated by forces arising from the relative displacement of the connected parts, in virtue of their elasticity.

The mechanical analogy for the electromagnetic field is explicated in terms of energy relations, rather than the pictorial mechanical model contrived in 'On physical lines of force'.

The abandonment of the model of molecular vortices and the cellular ether did not imply the rejection of mechanical foundations. He insists that the electromagnetic field is to be conceived as a 'complicated mechanism', embodied by an ether subject to energy transformations. 'Such a mechanism must be subject to the general laws of Dynamics' (*SP*, **1**: 533). He now deploys the Lagrangian formalism of abstract dynamics, based on the forces, velocities and displacements of a system of particles, which provides the dynamical basis for the statement of eight sets of 'General Equations of the Electromagnetic Field' (*SP*, **1**: 554–64).

He contrasts his new 'dynamical theory' with the argument of 'On physical lines of force'. He notes that 'on a former occasion' he had 'attempted to describe a particular kind of motion and a particular kind of strain, so arranged as to account for the phenomena'; here he aims 'to avoid any hypothesis of this kind'. The use of mechanical terminology in the present paper was to be considered 'as illustrative, not as explanatory'. The mechanism of molecular vortices and vortex cells was no longer to be regarded as providing a physical theory of lines of force and the electromagnetic field, but merely as illustrative, as a possible mechanical representation. In place of a mechanical model, described by vortex filaments or vortex cells, the electromagnetic field was to be embodied by energy transformations: 'In speaking of the Energy of the field, however, I wish to be understood literally'. Because 'all energy is the same as mechanical energy', whatever its form, the presence of energy in the electromagnetic field expresses the dynamical basis of the theory.

> On our theory [energy] . . . resides in the electromagnetic field, in the space surrounding the electrified and magnetic bodies, as well as in those bodies themselves, and is in two different forms, which may be described without hypothesis as magnetic polarization and electric polarization, or, according to a very probable hypothesis, as the motion and the strain of one and the same medium'. (*SP*, **1**: 563–4)

The transmission of energy in the electromagnetic field provides the connecting theme in Maxwell's expression of the dynamical basis of the theory. He maintains that the transmission of energy requires a medium in which it is propagated; and the essence of his field theory was the formulation of a 'dynamical theory' of this ethereal medium (see Chapter VIII.1). This argument is so fundamental to his whole field theory, that he chooses to elaborate it, in a concluding rhetorical flourish, in the final paragraph of his *Treatise on Electricity and Magnetism* (1873).

> In fact, whenever energy is transmitted from one body to another in time, there must be a medium or substance in which the energy exists after it leaves one body and before it reaches the other . . . and if we admit this medium as an hypothesis, I think it ought to occupy a prominent place in our investigations, and that we ought to construct a

> mental representation of all the details of its action, and this has been my constant aim in this treatise. (*Treatise*, **2**: 438 (§866))

In the *Treatise*, as in 'A dynamical theory of the electromagnetic field', this 'mental representation' of the 'medium' was explicated in the abstract form of general dynamical equations, rather than appealing to the concrete imagery of 'On physical lines of force'. He proceeded in terms of 'analytical mechanics' rather than 'physical mechanics' (the terms are Poisson's).[3]

Maxwell contrasts his two mechanical theories of the electromagnetic field in a letter to Peter Guthrie Tait in December 1867, seeking to correct Tait's version[4] of his intentions in the two papers.

> There is a difference between a vortex theory ascribed to Maxwell . . . and a dynamical theory of Electromagnetics by the same author in Phil Trans 1865. The former is built up to show that the phenomena are such as can be explained by mechanism. The nature of this mechanism is to the true mechanism what an orrery is to the Solar System. The latter is built on Lagranges Dynamical Equation and is not wise about vortices. (*LP*, **2**: 337)

The ether model was like a clockwork model of the solar system, an orrery, which could illustrate the order and motion of the planets. His 'theory of molecular vortices', he observed early in 1873, was 'rough and clumsy compared with the realities of nature, [but] may have served its turn as a provisional hypothesis' (*SP*, **2**: 306).

This is the position on the mechanical construction of an ether model that he affirms in the *Treatise*. There he makes the point that there could, in principle, be an infinite number of possible mechanical models which might be proposed to represent the electromagnetic field.

> The problem of determining the mechanism required to establish a given species of connexion between the motions of the parts of a system always admits of an infinite number of solutions. Of these, some may be more clumsy or more complex than others, but all must satisfy the conditions of mechanism in general.

This was a general condition of dynamical systems; hence the ether model of 'On physical lines of force' could be no more than illustrative, demonstrating the possibility of providing a mechanical explanation of electromagnetism.

> The attempt which I then made to imagine a working model of this mechanism must be taken for no more than it really is, a demonstration that mechanism may be imagined capable of producing a connexion

> mechanically equivalent to the actual connexion of the parts of the electromagnetic field. (*Treatise*, **2**: 416–17 (§831)

These remarks were made in discussing the Faraday magneto-optic effect, which provided the context for clarification of the status of the hypothesis of molecular vortices. The existence of molecular vortices, regarded as 'probable' in the second part of 'On physical lines of force' (*SP*, **1**: 468), was now described confidently as resting on 'good evidence', that of the magneto-optic effect, which, in the fourth part of 'On physical lines of force', had formed the capstone to the theory (see Chapter V.2).

Maxwell presents his discussion of the Faraday effect as 'an expansion' of Thomson's 1856 discussion of the Faraday magneto-optic rotation in terms of the rotation of molecular vortices, quoting from Thomson's paper at length (*Treatise*, **2**: 415–16 (§831)). The basic assumption in his argument favouring the existence of vortices, is the claim that a rotatory phenomenon must have a rotatory cause (see Chapter V.2). Writing to Thomson in January 1873, he observes that 'Faradays twist of polarized light will not come out without what the schoolmen call local motion' (*LP*, **2**: 784). Thus he claims that the Faraday effect is caused by the rotation of 'small portions of matter' connected by 'some kind of mechanism'.

> I think we have good evidence for the opinion that some phenomenon of rotation is going on in the magnetic field, that this rotation is performed by a great number of very small portions of matter, each rotating on its own axis, this axis being parallel to the direction of the magnetic force, and that the rotations of these different vortices are made to depend on one another by means of some kind of mechanism connecting them. (*Treatise*, **2**: 416 (§831))

While there was 'good evidence' for the existence of a 'mechanism' of matter in motion generating the rotation in the magnetic field, his own hypothetical 'working model', of rotating vortices and idle wheel particles, was merely illustrative of the possibility that a mechanism might be 'imagined'.

In the third part of 'On physical lines of force' Maxwell had formulated the concept of the 'displacement' of electricity in terms of his ether model of elastic vortex cells: the flow of electric 'charge' was represented by the accumulation of flowing particles (see Chapter V.2). This innovation formed a key element in the expression of his equations of electromagnetism. Yet his attempt, in 'A dynamical theory of the electromagnetic field', to emancipate his theory of 'displacement' from the ether model, led to difficulties in formulating the field equations. Detaching the concept of 'displacement' from the supposition of the accumulation of electric 'charge', led to inconsistencies of algebraic sign in the equations (*SP*, **1**: 554, 557, 561).[5] Subsequently he began to clarify the relation between 'charge' and 'displacement' (*SP*, **2**: 139). In a letter to Thomson in June 1869, giving an account of the development of his ideas as he was writing the *Treatise*, he mentioned a crucial new concept: he supposed that 'the movements of electricity are like those of an incompressible fluid' (*LP*, **2**: 486).

In the *Treatise* he considers electricity as a flowing incompressible fluid: 'at every instant as much must flow out of any given closed space as flows into it' (*Treatise*, **1**: 64 (§62)). The fluid does not consist of electric charges; rather 'all electrification is the residual effect of polarization of the dielectric'. Electric 'charge' is conceived as the manifestation of the flow of electricity, not as an accumulation of electric particles; and charge only becomes apparent at the boundary of the dielectric:

> polarization exists throughout the interior of dielectrics, but it is there neutralized by the juxtaposition of oppositely electrified parts, so that it is only at the surface of the dielectric that the effects of the electrification become apparent. (*Treatise*, **1**: 133 (§111))

This theory provides a coherent representation of electric charge in terms of his theory of the primacy of the electromagnetic field: charge is a manifestation of the field.[6]

The mathematical argument of the *Treatise* encompasses a development of the generalised Lagrangian theory of the electromagnetic field as presented in 'A dynamical theory of the electromagnetic field'. Maxwell's correspondence with Tait indicates that his revision of his dynamical method, into the form as presented in the published text of the *Treatise*, began around May 1872 (*LP*, **2**: 716). In a postcard in June 1872 he wrote that

> I have been overhauling the Equations of motion and have got a way of deducing them (in Hamilton's form) from the variables their velocities and the forces acting *on them* alone . . . by beginning with impulsive force. (*LP*, **2**: 732–3)

The method he follows in his chapter 'On the equations of motion of a connected system' (*Treatise*, **2**: 184–94 (§§553–67)), was developed from that adopted by Thomson and Tait in their *Treatise on Natural Philosophy* (1867), where the generalised equations of motion are derived from impulsive forces.

The fundamental mathematical axiom in Thomson and Tait's presentation of dynamics is based on a minimum theorem formulated by Thomson in 1849: that the energy of motion generated in a liquid of finite dimensions is less than the energy of any other motion consistent with the same motion of its bounding surface.[7] Thomson generalised this theorem in 1863, stating a general minimum property of a material system, which Tait termed 'Laziness'.[8] The theorem was stated and proved in *Natural Philosophy*: that the variation of a system by impulsive forces is such that the kinetic energy is a minimum.[9] Because an impulsive force acts in an infinitesimal time-increment, the configuration and potential energy of the system would be unaltered; only the kinetic energy of the system would vary. (They introduced the term 'kinetic energy'.) Thomson and Tait derived generalised (Lagrangian) equations of motion from the supposition of impulsive forces.

In seeking to develop his dynamical theory of the electromagnetic field, Maxwell sought to join mathematical language to physical understanding. He propounds his style of reasoning in mathematical physics in his review of Thomson and Tait's *Elements of Natural Philosophy* (1873). He urges that a purely symbolic theory, employing 'the machinery with which mathematicians have been accustomed to work problems about pure quantities', was inadequate; to be intelligible, the mathematical formalism had to be 'clothed with the imagery . . . of the phenomena of the science itself' (*SP*, **2**: 325). In mathematical physics, any symbolic representation had to provide a physical interpretation of nature.

In the *Treatise*, he stresses his aim of formulating a dynamical theory of the field that would highlight the physical meaning of the dynamical equations. He contrasts Lagrange's method, which he considers to be a formalism of generalised equations of motion conceived as 'pure algebraical quantities', in a manner 'free from the intrusion of dynamical ideas', with the method by which, in their *Treatise on Natural Philosophy*, Thomson and Tait had sought to 'cultivate our dynamical ideas'. Whereas the Lagrangian method provides a mathematical formalism, which avoids reference to the concepts of velocity, momentum and energy after they had been replaced by symbols in the equations of motion, Thomson and Tait had placed emphasis on the physical meaning of the dynamical concepts. Their method, deriving generalised equations of motion from impulsive forces, 'kept out of view the mechanism by which the parts of the system are connected', and thus satisfied his overarching aim of avoiding the formulation of a specific ether model. But at the same time, their method did keep 'constantly in mind the ideas appropriate to the fundamental science of dynamics'. Thomson and Tait's method satisfied Maxwell's criterion of providing a dynamical explanation (*Treatise*, **2**: 184–5, 193–4 (§§554, 567)).

In one of the concluding sections of the *Treatise*, in a final exhortation justifying his adoption of field theory in preference to an action at a distance theory of electromagnetism, Maxwell cites a letter from Carl Friedrich Gauss to Wilhelm Weber, written in 1845 and published in 1867 in the fifth volume of Gauss' collected works. Not surprisingly, he found a vindication of his own approach in Gauss' remarks (see Chapter VIII.1).

> In a very interesting letter of Gauss to W. Weber he refers to the electrodynamic speculations with which he had been occupied long before, and which he would have published if he could then have established that which he considered the real keystone of electrodynamics, namely, the deduction of the force acting between electric particles in motion from the consideration of an action between them, not instantaneous, but propagated in time, in a similar manner to that of light. He had not succeeded in making this deduction when he gave up his electrodynamic researches, and he had a subjective conviction that it would be necessary in the first place to form a consistent representation of the manner in which the propagation takes place.
>
> (*Treatise*, **2**: 435 (§861))

Tracing the transformations of energy, the bedrock of his dynamical theory of the field, was fundamental to his aim of formulating a 'consistent representation', his rendition of Gauss' term '*construirbare Vorstellung* [constructible representation]'.[10] The construction, by a dynamical theory of the transmission of energy, of a 'mental representation' of the medium propagating electromagnetic action had, he affirms, been his 'constant aim in this treatise' (*Treatise*, **2**: 438 (§866)). This dynamical theory was *consistent* with the requirement of intelligibility: the expression of 'dynamical ideas from a physical point of view'. In Maxwell's dynamical theory there is a link between the mathematical formalism and the physical reality depicted; 'we must', he writes, 'have our minds imbued with these dynamical truths as well as with mathematical methods' (*Treatise*, **2**: 184, 194 (§§554, 567)).

As he indicated to Tait in June 1872, in his exposition of the equations of analytical dynamics in the *Treatise* Maxwell favours the expression of kinetic energy adopted in William Rowan Hamilton's memoir 'General methods in dynamics' (1834–5). In this method, the kinetic energy is expressed in terms of the variables (q) and momenta (p) of the particles constituting a material system. Thus $q = \mathrm{d}T_p/\mathrm{d}p$, where, in Maxwell's words, 'the velocity corresponding to the variable q is the differential coefficient of T_p [the kinetic energy] with respect to the corresponding momentum p'. The kinetic energy of the system changes as a result of the action of an infinitesimal force impulse, where $p = \int F\,\mathrm{d}t$; but the instantaneous state of the system is independent of the cause of motion. Consistent with the aims of his dynamical theory of electromagnetism, the hidden machinery of the system is not specified: 'the variables, and the corresponding velocities and momenta, depend on the actual state of the motion of the system at the given instant, and not on its previous history'. He is therefore able to obtain Hamilton's equation of motion for the impressed forces acting on a material system, without reference to the forces which established its configuration:

$$F_r = \frac{\mathrm{d}p_r}{\mathrm{d}t} + \frac{\mathrm{d}T_p}{\mathrm{d}q_r}$$

where the momentum p_r and the force F_r belong to the variable q_r (*Treatise*, **2**: 186–90 §§556–61)).

In a draft he explains his preference for the form of the equation of motion given by Hamilton over that given by Lagrange. This is because of the physical meaning of the Hamiltonian form of the equation, which is based on momentum rather than (as with Lagrange) on velocity.

> Now it is not the velocities which obey Newton's law of persevering in their actual state, but the momenta or 'quantities of motion'. Hence if we wish to apply Newton's law we must express the kinetic energy in terms of the momenta and use Hamilton's form of the equations of motion. (*LP*, **2**: 746)

In emphasising the physical basis of dynamics, and Newton's laws of motion, Maxwell echoes the framework of Cambridge dynamics as taught by William Hopkins (see Chapter II.2). In the *Treatise*, he observes that Lagrange had sought, in his *Mécanique Analytique*, 'not only to dispense with diagrams, but even to get rid of the ideas of velocity, momentum and energy, after they have been once for all supplanted by symbols in the original equations'. By contrast, in Maxwell's method, the 'language of dynamics' would express 'some property of moving bodies' (*Treatise*, **2**: 185, 194 (§§554, 567)). The physical meaning of Hamilton's equation of motion is rendered intelligible by its expression of the dynamical concept of momentum; and hence Newton's second law of motion would determine the meaning of the dynamical theory of the electromagnetic field.

These remarks were prompted by his interpretation of Thomson and Tait's exposition of dynamical theory. In a review of the second edition of their *Natural Philosophy* in 1879, he further clarifies their dynamical method. He likens the generalised dynamical method to bellringers pulling on ropes in a belfry. The machinery in the belfry could be imagined as complex, but 'all this machinery is silent and utterly unknown to the men at the ropes'. By tugging on the ropes the bellringers can give them any position and velocity.

> These data are sufficient to determine the motion of every one of the ropes when it and all the others are acted on by any given forces. This is all that the men at the ropes can ever know.

This was the implication of the method of Thomson of Glasgow and Tait of Edinburgh, whom he describes as the 'northern wizards' (making allusion to the 'Wizard of the North', Sir Walter Scott). The hidden machinery of the dynamical system of the electromagnetic field would therefore be 'silent and utterly unknown' (*SP*, **2**: 782–4).

VI.2 Statistical physics

On writing to Stokes in May 1859, about his new interest in the kinetic theory of gases, Maxwell placed emphasis on the cardinal feature of his method.

> Of course my particles have not all the same velocity, but the velocities are distributed according to the same formula as the errors are distributed in the theory of 'least squares'. (*LP*, **1**: 610)

The statistical distribution of the velocities of the gas molecules was a distinct conceptual advance on Clausius' probabilistic reasoning. Treating the motion of the gas molecules as a random process, Clausius had recognised that there would be considerable variation in the velocities of the molecules; but in his mathematical analysis of molecular encounters, he simply used the average molecular velocity. Maxwell, however, maintained that a statistical analysis, analogous in form to Laplace's representation of the distribution of errors, was required: he suggested that the velocities were distributed among the molecules in accordance with a statistical distribution function.

Clausius' discussion of the probabilities of molecular encounters would in itself have very likely attracted Maxwell's attention, regardless of other points of interest in his paper (see Chapter V.1). He had been interested in the theory of probability as early as 1850.

> They say that Understanding ought to work by the rules of right reason. These rules are, or ought to be, contained in Logic; but the actual science of Logic is conversant at present only with things either certain, impossible, or *entirely* doubtful, none of which (fortunately) we have to reason on. Therefore the true Logic for this world is the Calculus of Probabilities, which takes account of the magnitude of the probability (which is, or which ought to be in a reasonable man's mind). (*LP*, **1**: 197)

These remarks, addressed to his friend Lewis Campbell, and probably dating from July 1850, may have been stimulated by reading an essay by Sir John Herschel, on Adolphe Quetelet's *Theory of Probabilities*, in the *Edinburgh Review* of July 1850.[11] In 1849–50, when Maxwell was his student at Edinburgh University, J. D. Forbes was embroiled in a debate with Herschel over the use of probability arguments to determine the distribution of double stars. Around 1850 the logic of probabilities was widely debated in Britain: both John Stuart Mill, in his *System of Logic* (1843) and George Boole, in his *Investigation of the Laws of Thought* (1854) – authors read by Maxwell – had participated in this debate.[12]

Maxwell's acquaintance with the ideas of social statistics probably owed much to his reading of Henry Thomas Buckle's *History of Civilization in England* (1857–61); he read the first of its two volumes on its publication in 1857, describing it to Lewis Campbell as a 'bumptious book', but containing 'a great deal of actually original matter, the true result of fertile study' (*LP*, **1**: 576). Buckle applied Quetelet's principle of statistical regularity, the stability of the mean, into a deterministic reading of history. Regular social laws, rather than individual contingency, were held to shape historical development. Maxwell found Buckle's determinism problematic (see Chapter IX.2); but it is likely that his interest in contemporary debates on the nature and implications of statistical reasoning, made him receptive to Clausius' introduction of a statistical argument in discussing the collision of particles.

Whatever stimulus may have come from Quetelet's social statistics and notion of the 'average man', Maxwell's statistical theory did not rest on the use of mean values, or on the certainty of the collective averages of social statistics. He was interested in grasping the significance of the error distribution function in understanding deviations from mean values.[13] Reading Herschel's review of Quetelet, which he would have encountered when it was reprinted in Herschel's collected *Essays*, which he read in the winter of 1857–8 (*LP*, **1**: 576, 583), probably did help to shape his statistical argument. There are formal similarities between his derivation of the distribution law in 'Illustrations of the dynamical theory of gases' (1860) and Herschel's discussion of the error function.[14]

In discussing the collisions of elastic particles in this first derivation of his distribution function, he seeks to determine 'the average number of particles whose velocities lie between given limits' after collisions have occurred. He makes two assumptions, both of which have parallels in Herschel's account of the error law. First, he assumes that the velocity components along each of the directions in space (x, y, z) are 'independent' of the others; second, that the density of the distribution is spherically symmetric. The distribution is therefore the product of three independent terms and 'the distance from the origin'. Hence

> the velocities are distributed among the particles according to the same law as the errors are distributed among the observations in the theory of the 'method of least squares'. (*SP*, **1**: 380–2)

He introduced a new argument, revising his derivation of the distribution function, in his Royal Society paper 'On the dynamical theory of gases' (1867). He explains, that since

> the assumption that the probability of a molecule having a velocity resolved parallel to x lying between given limits is not in any way affected by the knowledge that the molecule has a given velocity resolved parallel to y . . . may appear precarious (*SP*, **2**: 43)

he now presents a different derivation. He demonstrates that the velocity distribution would maintain a state of equilibrium, and calculates the effect of collisions on the distribution law; the equilibrium distribution is the distribution that would be unchanged by collisions.

This revision of his argument was prompted by a broader reconstruction of his kinetic theory of gases (see Chapter VIII.2). His discovery that the viscosity of gases was a linear function of the absolute temperature was not compatible with the hypothesis, adopted in 'Illustrations of the dynamical theory of gases', that gas molecules could be represented as colliding elastic spheres. He therefore abandons this model in his paper 'On the dynamical theory of gases'. The shift from a 'physical analogy' of colliding elastic particles, stripping away the physical model to represent gas molecules as centres of force subject to a law of repulsion, parallels the transition in his field theory between 1861 and 1865, from the ether model of 'On physical lines of force' to the nonhypothetical stance of 'A dynamical theory of the electromagnetic field'. 'The molecules of a gas in this theory', he now writes, 'are those portions of it which move about as a single body' (*SP*, **2**: 33). He computes the motions of molecules travelling in complicated trajectories, using the methods of orbital dynamics (*SP*, **2**: 40; *LP*, **2**: 256). He finds that if the force law between the molecules was 'inversely as the *fifth* power of the distance', the viscosity varied directly with the absolute temperature, consonant with his experiments (*SP*, **2**: 29, 41–2, 71).

He considered the theory to be a 'dynamical' theory of gases, because he supposes gases to 'consist of molecules in motion'. Each molecule is conceived as 'a small body consisting in general of parts capable of being set into various kinds of motion'; and 'the action between the molecules is supposed to be that of bodies repelling each other at a distance . . . the law of force [being] . . . that of the inverse fifth power of the distance' (*LP*, **2**: 279–80). Writing to Stokes, as Secretary of the Royal Society, in December 1866, he explains his usage: 'I therefore call the theory a dynamical theory because it considers the motions of bodies as produced by certain forces' (*LP*, **2**: 291). This defines the dynamical basis of his theory of gases.

There is no suggestion, in Maxwell's formulation of his 'dynamical' theory of gases, of any disharmony between the dynamical basis of the theory, as supposing a system of molecules acted upon by forces, and the statistical distribution of molecular velocities, which forms the basis of the theory of molecular encounters and the explanation of the viscosity, diffusion and thermal conductivity of gases. But in his introductory lecture at Cambridge in October 1871, he contrasts 'strict dynamical principles' with the inherently uncertain knowledge generated by the 'statistical method'.

> [I]n applying dynamical principles to the motion of immense numbers of atoms, the limitation of our faculties forces us to abandon the attempt to express the exact history of each atom, and to be content with estimating the average condition of a group of atoms large enough to be visible. This method of dealing with groups of atoms, which I may call the statistical method, and which in the present state of our knowledge is the only available method of studying the properties of real bodies, involves an abandonment of strict dynamical principles, and an adoption of the mathematical methods belonging to the theory of probability. (*SP*, **2**: 253)

He thus emphasises the limitations of statistical knowledge, the incompleteness of explanation generated by the statistical method. He points to a disjunction between the certain predictions of dynamical principles, which could trace the paths of individual particles, and the reliance on probabilities forced on the physicist dealing with an immense number of particles, their motion being represented by a statistical function.

He had first introduced the statistical method of his kinetic theory of gases as a means of describing the complex pattern of molecular collisions (see Chapter V.1). He had been unable to compute the trajectories of the particles constituting the rings of Saturn 'with any distinctness' (*SP*, **1**: 354), and statistical analysis provided a method of calculation. His distinction between dynamical and statistical principles highlights the contrast between tracing the 'orbit & excentricity' of the particles of Saturn's rings (*LP*, **1**: 555), and the statistical theorems, expressing the regularity and equilibrium of the motion of molecules, developed in his 'Kinetic theory of gases', as he termed the theory in 1871 (*LP*, **2**: 654).

But there was an additional component in his reasoning in 1871. The terms used in the passage cited above from his Cambridge inaugural lecture, echo remarks which appear in one of the final paragraphs of his text the *Theory of Heat*, published in the same year.

> In dealing with masses of matter, while we do not perceive the individual molecules, we are compelled to adopt what I have described as the statistical method of calculation, and to abandon the strict dynamical method in which we follow every motion by the calculus.

This statement appears in the penultimate section of the book, which bears the heading 'Limitation of the second law of thermodynamics'. Here he argues that because of the statistical distribution of molecular velocities in a gas at equilibrium, the process of heat flow from a hot to a colder body (in accordance with the second law of thermodynamics) could be abrogated at the molecular level. Spontaneous fluctuations of individual molecules could occur, fluctuations which could take heat from a cold body to a hotter one. The second law of thermodynamics, in Maxwell's interpretation, is therefore a statistical law, which applies to systems of molecules, not to the motions of individual molecules. Moreover, he contrasts the essential irreversibility of natural processes, as described by the second law of thermodynamics, and the perfect reversibility of the motions of particles as described by the laws of dynamics (see Chapter VI.3). These reflections help to shape his articulation of the distinction between the dynamical and statistical *methods*.

The statistical theory of gases established the context in which he draws these distinctions. He makes this plain in an earlier section of the *Theory of Heat*, 'On the kinetic theory of gases'. Here he points the contrast between the 'statistical method', which generates knowledge of molecules in the mass, as distinct from the 'kinetic method' (as he here terms it), which traces the path of an individual molecule.

> I wish to point out that, in adopting this statistical method of considering the average number of groups of molecules selected according to their velocities, we have abandoned the strict kinetic method of tracing the exact circumstances of each individual molecule in all its encounters. It is therefore possible that we may arrive at results which, though they fairly represent the facts as long as we are supposed to deal with a gas in mass, would cease to be applicable if our faculties and instruments were so sharpened that we could detect and lay hold of each molecule and trace it through all its course.

He pursues this argument in his discussion of the 'Limitation of the second law of thermodynamics', which follows shortly in his text. Here he introduces a 'being' (that is: 'Maxwell's demon'), with 'faculties so sharpened that he can follow every molecule in its course'.[15] It would require the action of such a 'being' to select individual molecules, and to produce an observable flow of heat from a cold body to a hotter one, and hence violate the second law of thermodynamics (see Chapter VI.3). His contrast between tracing the trajectory of a molecule, by the 'dynamical method', and considering groups of molecules selected according to their velocities, by the 'statistical method', thus bears on his reflections on the meaning of the second law of thermodynamics as an essentially statistical law. Different methods, dynamical and statistical, were therefore appropriate in tracing 'every molecule in its course' and representing 'a gas in mass'.

These remarks, on the distinction between dynamical and statistical methods, need to be set alongside his discussion of dynamical explanation in the *Treatise on Electricity and Magnetism*. There he expounds his project of developing a dynamical theory of the electromagnetic field. He affirms the necessity of keeping 'constantly in mind the ideas appropriate to the fundamental science of dynamics'; this was basic to his aim of constructing a 'consistent representation' of the medium in which energy is transmitted in the field (*Treatise*, **2**: 194, 435 (§§567,861)). The terms 'dynamical' or 'kinetic', as applied to the theory of gases, indicate that a gas is comprised of a mass of molecules in motion, and that these motions are produced by the action of forces. The language he uses in expounding his theory of gases continued to highlight the 'dynamical' assumptions and methods of the theory.

In his lecture to the Chemical Society in February 1875 'On the dynamical evidence of the molecular constitution of bodies', he begins by affirming that

> when a physical phenomenon can be completely described as a change in the configuration and motion of a material system, the dynamical explanation of that phenomenon is said to be complete. We cannot conceive any further explanation to be either necessary, desirable, or possible .... (*SP*, **2**: 418)

This assertion echoes the discussion of dynamical theory in the *Treatise*. Its statement in the context of a lecture on the theory of gases and molecules, makes clear the generality of dynamical theory in physics. The theory of the constitution of bodies should therefore be based on a 'dynamical method of studying a material system consisting of an immense number of particles'. This is Maxwell's kinetic theory of gases. By forming an idea of the 'configuration and motion, and of the forces acting on the particles', it is then possible to deduce, from the dynamical theory, empirical relations, such as Boyle's law of gases, 'which are capable of being observed in visible portions of the system' (*SP*, **2**: 420).

Throughout the lecture he seeks to show that the kinetic theory of gases, a theory of molecules in motion acted on by forces, gave evidence of the application of 'dynamical methods to chemical science'; that the kinetic theory of gases, without reference to a statistical analysis of molecular motions, could establish chemical doctrines such as Avogadro's hypothesis (see Chapter V.1). In illustration, he gives an account of Clausius' concept of the 'virial', which he describes as a 'new dynamical idea', a function which relates the kinetic energy of a system of molecules to the coordinates of the particles and the forces acting on them. The 'virial' enabled laws of gases to be deduced from dynamical principles, and it provided insight into the properties of matter (see Chapter VIII.2). The application of these 'dynamical' concepts did not require that the trajectories of individual particles be traced. But where it was necessary to consider molecular collisions, then the statistical method, where molecules are distributed into groups according to their velocities, had to be employed. The statistical method 'is the only one available either experimentally or mathematically' for treating gases as 'moving systems of innumerable molecules'. He declares that, by invoking the statistical method, 'we pass from the methods of strict dynamics to those of statistics and probability'. By the 'methods of strict dynamics' he refers to 'the impossible task of following every individual molecule through all its encounters'. This attempt to trace the trajectory of individual molecules was therefore to be abandoned, in favour of the statistical method deployed in his kinetic theory of gases: the task 'of registering the increase or decrease of the number of molecules in the different groups' (*SP*, **2**: 421–9).

In his 1873 lecture on 'Molecules', Maxwell again draws the distinction between the statistical and the dynamical/kinetic methods, introducing another terminological variant for the 'dynamical method': the 'historical method'. This term avoids any ambiguity which might arise from use of the word 'dynamical'. The distinction between 'statistical' and 'historical' methods highlights the contrast between considering a group of molecules selected according to their velocities, and tracing the path of an individual molecule through time.

> The equations of dynamics completely express the laws of the historical method as applied to matter, but the application of these equations implies a perfect knowledge of all the data.

The task of tracing the trajectory of individual molecules, so as to obtain a perfect knowledge of all the data of a material system, was not possible in studying the properties of a mass of a gas.

> But the smallest portion of matter which we can subject to experiment consists of millions of molecules, not one of which ever becomes individually sensible to us. We cannot, therefore, ascertain the actual motion of any one of these molecules; so that we are obliged to abandon the strict historical method, and to adopt the statistical method of dealing with large groups of molecules.

Once again he emphasises the limitations of statistical knowledge: the 'regularity of averages' is substituted for the 'absolute precision which belongs to the laws of abstract dynamics' (*SP*, **2**: 374).

In a draft of his lecture he reflects on the different forms of knowledge generated by the application of the two methods. The shift from the historical or dynamical method to the statistical method, he notes, is a 'step the philosophical importance of which cannot be overestimated', a step 'equivalent to the change from absolute certainty to high probability'. Despite the 'stability of the averages of large numbers of variable events . . . in a particular case a very different event might occur' from that expected from the 'regularity of averages'; we are, nevertheless, 'morally certain that such an event will not take place' (*LP*, **2**: 930–3).

VI Dynamical and statistical explanation

The philosophical doctrine of moral certainty was conventionally used to justify reasoning in cases where absolute certainty, as in cases such as identical propositions or geometrical demonstration, could not be attained. Moral certainty was construed as probable certainty, the certainty which would be accepted by reasonable persons. The distinction between absolute and moral certainty was used to establish the rationality of a limited kind of certainty. In seeking to justify the validity of moral certainty against Hume's sceptical assault on the reasoning process involved in probable judgements, Scottish philosophers appealed to common sense, claiming that the certainty allowed in ordinary life could legitimise judgements based on probability, which were regarded as being morally certain.[16] Maxwell uses the distinction between absolute and moral certainty to highlight the limitation but also the validity of his statistical reasoning. The 'statistical method' is based on probabilistic reasoning and therefore could be conceived as providing 'moral' certainty, whereas the 'dynamical method' could be regarded as having 'absolute' predictive power. Maxwell thus sought to expound the implications of statistical reasoning in physics, to explain his reasons for abandoning the 'dynamical' or 'historical' method, and to justify the form of knowledge generated by the 'statistical' method.

In a review in the journal *Nature* in 1877, Maxwell further clarified his distinction between the dynamical and statistical methods. To explain the reasoning involved in the statistical method, the method appropriate to 'dealing with the fluctuations of a large population' of molecules, he introduces the analogy of a turnstile. A turnstile counts individuals without respect to their previous history or identity, just as in the statistical method deployed in the kinetic theory of gases no attempt is made to follow the history and trajectory of individual molecules.

> We divide the bodies of the system into groups according to their position, their velocity, or any other property belonging to them, and we fix our attention not on the bodies themselves, but on the *number* belonging at any instant to one particular group. This number is, of course, subject to change on account of bodies entering or leaving the group, and we have therefore to study the conditions under which bodies enter or leave the group, and in so doing we must follow the course of the bodies according to the dynamical method. But as soon as the process is over, when the body has fairly entered the group or left it, we withdraw our attention from the body, and if it should come before

> us again we treat it as a new body, just as the turnstile at an exhibition counts the visitors who enter without respect to what they have done or are going to do, or whether they have passed through the turnstile before.[17]

These remarks were made in a review of H. W. Watson's *Treatise on the Kinetic Theory of Gases* (1876), where a theorem of Ludwig Boltzmann's is discussed. The theorem states that a system in equilibrium will, if undisturbed, pass through every mechanical state that is compatible with its total energy. In 1878, in a major paper on statistical mechanics, Maxwell enlarged on Boltzmann's work. To produce this property (later called 'ergodic'), he assumes that the particles encounter disturbances, such as the collision of gas molecules with the vessel containing them.

As in his 'dynamical theory of gases', he provides both dynamical and statistical representations of the system of particles. In his 'Dynamical Specification of the motion' he employs Hamilton's equations of motion, the method which he had used in the *Treatise* (see Chapter VI.1). He describes the configuration of the system in terms of variables of position and momentum, specifying the system in a particular 'phase of the motion of the system', the phases through which the system passes being 'consistent with the equation of energy'. This is a generalisation of his dynamical theory of the motion of particles as applied to the theory of gases.

To connect this dynamical representation to the 'Statistical Specification' he employs a 'wider definition' of the 'phase' of a system.

> In the statistical method of investigation, we do not follow the system during its motion, but we fix our attention on a particular phase, and ascertain whether the system is in that phase or not, and also when it enters the phase and when it leaves it.

He introduces the concept of an ensemble of systems 'similar to each other in all respects except in the initial circumstances of the motion . . . the total energy being the same in all' (*SP*, **2**: 715–17, 720–21). He examines the statistical distribution of an ensemble of systems in a given state at any instant of time, rather than (as in his previous analyses) considering the distribution of a single system over time. The method of ensembles generalised and reinforced his view of statistical representation as reflecting the limitation in knowledge of systems of particles.[18]

In March 1879 Maxwell reviewed, for the Royal Society, a paper by Osborne Reynolds on the theory of gases. In this substantial review, whose course of argument was stimulated by his current interest in rarefied gas dynamics (see

Chapter VIII.2), he further elucidates the cardinal feature of his method which he had illustrated by the analogy of a turnstile: the avoidance of considering the history and trajectory of individual molecules.

> The method which I have used . . . contrives to avoid the consideration of the places from which the molecules have come . . . . But I carefully abstain from asking the molecules which enter [an element of space] where they had started from, I only count them and register their mean velocities, avoiding all personal enquiries which would only get me into trouble.[19]

VI.3 The 'demon' and the second law of thermodynamics

In December 1867, in response to a query from his friend Peter Guthrie Tait, asking whether he was 'sufficiently up to the history of thermodynamics to critically examine & put right a little treatise I am about to print', Maxwell introduced one of his most famous ideas, that of the 'demon'. The book to which Tait was alluding, and on which he was requesting Maxwell's critical advice, was his *Sketch of Thermodynamics* (1868), an enlarged version of two essays, on 'The dynamical theory of heat' and on 'Energy', which he had published in the *North British Review* in 1864. Tait had developed these essays into two draft chapters of his book, which were privately printed and circulated to interested parties (including Helmholtz, Rankine and Clausius). As he ruefully confessed to Maxwell, 'Clausius & others have cut up very rough about bits referring to them'; adding, rather disingenuously, that 'I dont pretend to know the subject thoroughly and would be glad of your help'.[20]

In reply, Maxwell warned Tait that he did 'not know in a controversial manner the history of thermodynamics, that is I could make no assertions about the priority of authors without referring to their actual works'. Tait's text did little to dispel his state of ignorance, though over the next few years he grappled with the sources, and ultimately clarified his historical understanding. But in response to Tait's request for advice, he expressed himself willing to assist so far as he could: 'any contributions I could make to that study are in the way of altering the point of view here & there for clearness or variety and picking holes here & there to ensure strength & stability' (*LP*, **2**: 330–1). This proved to be a characteristically modest disclaimer, for the hole he picked bored in to the very foundations of the subject: the bearing of his statistical theory of the distribution of velocities among gas molecules on the interpretation of the second law of thermodynamics.

Maxwell's serious engagement with the science of thermodynamics, which led him to write his text on the *Theory of Heat* (1871), emerged as a by-product of his work on the kinetic theory of gases. He had, of course, long been aware of Thomson's work on the subject. In May 1855 he had asked Thomson, who, with Clausius, had formulated the second law of thermodynamics as an expression of the direction of heat flow, 'do you profess to account for what becomes of the vis viva [kinetic energy] of heat when it passes thro' a conductor from hot to cold?' (*LP*, **1**: 307). Thomson himself had raised a similar query in 1849,[21] which he had answered by clarifying the concept of irreversibility.

In 1850 Clausius had set out the framework of the science of thermodynamics, stating two fundamental principles: the equivalence of heat and work (the principle of the conservation of energy); and the principle explaining the generation of work from heat in a cyclic process, according to which, in falling between two temperature levels, part of the heat is converted into work while the rest descends to the lower temperature. These two principles became known as the two laws of thermodynamics. In his paper 'On the dynamical theory of heat' (1851) Thomson outlined the scope, as he understood it, of the conceptual problems facing the emergent science of 'thermodynamics', a term that he was himself to coin in 1854.[22] He placed emphasis on the problem of irreversibility in explaining the direction of heat flow.

Thomson's 'dynamical' theory of heat asserts that heat consists in the motion of the particles of bodies. He maintains that the mechanical work, which might have been generated from the heat expended in conduction through a solid body, has been wasted but not destroyed; the heat is transformed into the energy of motion of the particles of the body. Though the heat is unrecoverable, it has been transformed and dissipated: 'When heat is diffused by *conduction*, there is a *dissipation* of mechanical energy, and perfect *restoration* is impossible'. This is held to be a 'necessary' consequence of the second law of thermodynamics, stated in the following words:

> It is impossible by means of inanimate material agency, to derive mechanical effect from any portion of matter by cooling it below the temperature of the coldest of the surrounding objects.

While there was 'an absolute waste of mechanical energy' in irreversible processes, the heat dissipated in conduction is not destroyed: there 'must be some transformation of energy' into the energy of motion of the particles of the conductor. Thomson drew a general conclusion: 'There is at present in the material world a universal tendency to the dissipation of mechanical energy'.[23]

In drafting his paper 'On the dynamical theory of gases' early in 1866, Maxwell considered the question of the equilibrium of temperature in a vertical column of gas under gravity. He found that, according to his theory, the temperature of the gas would diminish as the height increases, at a greater rate than it does by expansion when a portion of gas is carried up bodily. Hence, 'the condition of final equilibrium of heat in a gas acted on by gravity is one of mechanical instability'. From this he drew the conclusion that energy could be obtained from a gas while the gas cooled.

> If however the motion were properly regulated the energy thus developed could be transferred to machinery so as to convert the invisible agitation of the gas into any other form of energy and thus form a perpetual motion. . . . Thus from a mass of gas acted on by gravity energy may be abstracted to any amount, and the gas cooled to a corresponding extent.

This apparent consequence of the theory of gases, that energy could be obtained from a cooling gas, was in conflict with the second law of thermodynamics.

> This result is directly opposed to the second law of Thermodynamics which affirms that 'it is impossible by means of inanimate material agency to derive mechanical effect from any portion of matter by cooling it below the temperature of the coldest part of the surrounding objects' (*LP*, **2**: 265–6)

quoting Thomson's statement of the law.

Before sending the paper to the Royal Society, he wrote to Thomson in February 1866, seeking advice about this problematic and unwelcome consequence of his dynamical theory of gases. He emphasised the significance of his conclusion, explicitly evoking the language of Thomson's own statement of the second law of thermodynamics:

> by means of material agency mechanical effect is derived from the gas under gravity by cooling it below the temperature of the coldest of the surrounding objects. See Thomson Dyn Θ of H 2nd Law.

The conclusion he drew would, if correct, have far-reaching consequences for the kinetic theory of gases: that 'there remains as far as I can see a collision between Dynamics & thermodynamics', a contradiction between the argument of the dynamical theory of gases and the second law of thermodynamics (*LP*, **2**: 269).

By the time he submitted the paper to the Royal Society in May 1866, he had discovered one of the errors in his analysis, a mistake in mathematical reasoning. But this partial correction of the argument led to the conclusion that the temperature of the gas would increase with height. He notes that while this did 'not lead to mechanical instability', it was 'equally inconsistent with the second law of thermodynamics' (*SP*, **2**: 75–6; *LP*, **2**: 283–4). Thomson reviewed the paper for the Royal Society, but was unable to detect the error that still compromised its argument: 'What the flaw may be in Maxwell's investigation if any I have not been able to see'.[24] However, in December 1866 Maxwell reported to Stokes (as Secretary of the Royal Society) that he had located the error in his statistical reasoning: 'I now make the temperature the same throughout' (*LP*, **2**: 292). He now found that a difference in the number of molecules projected upwards and downwards from the same stratum in the column of gas, would counteract the tendency to a transmission of energy. This was incorporated in an 'addition' which was appended to the paper prior to its publication in the *Philosophical Transactions*, a result which was consistent with the second law of thermodynamics and with 'the law of distribution of velocities to which we were led by independent considerations' (*SP*, **2**: 75–6).

On reviewing the issue in 1873, and explaining the development of the argument as published in his paper, he confessed that his initial conclusion, which implied a contradiction between his dynamical theory of gases and thermodynamics, had 'nearly upset my belief in calculation' (*LP*, **2**: 854). Even though he had resolved this 'paradox', as he had described it to Thomson (*LP*, **2**: 267, 269), it was nevertheless an issue he continued to consider.[25] Ruminating on the equilibrium of temperature in a column of gas under gravity, even though he had satisfactorily dispatched the difficulty, exposed the more general question, of the bearing of his statistical theory of the distribution of velocities among gas molecules on the interpretation of the second law of thermodynamics. It is likely that he had continued to reflect on the issue, for Tait's letter of December 1867 immediately prompted him to set down his ideas on this subject. Even before seeing the draft chapters of Tait's book on thermodynamics (*LP*, **2**: 335), on which his advice was sought, he raised a problem of his own for Tait to consider.

Maxwell told Tait that he would 'pick a hole' in the second law of thermodynamics, by exposing the problem of explaining the irreversible flow of heat from warmer to colder bodies in terms of the statistical regularities which describe the motion of gas molecules. He explains the issue as follows.

> Now let *A* & *B* be two vessels divided by a diaphragm and let them contain elastic molecules in a state of agitation which strike each other and the sides. Let the number of particles be equal in *A* & *B* but let those in *A* have the greatest energy of motion. . . . I have shown that there will be velocities of all magnitudes in *A* and the same in *B* only the sum of the square of the velocities is greater in *A* than in *B*. When a molecule is . . . allowed to go through a hole in the [diaphragm] . . . no work would be lost or gained, only its energy would be transferred from the one vessel to the other.
>
> Now conceive a finite being who knows the paths and velocities of all the molecules by simple inspection but who can do no work, except to open and close a hole in the diaphragm, by means of a slide without mass. Let him first observe the molecules in *A* and when he sees one coming the square of whose velocity is less than the mean sq. vel. of the molecules in *B* let him open the hole & let it go into *B*. Next let him watch for a molecule in *B* the square of whose velocity is greater than the mean sq. vel. in *A* and when it comes to the hole let him draw the slide & let it go into *A*, keeping the slide shut for all other molecules.
>
> Then the number of molecules in *A* & *B* are the same as at first but the energy in *A* is increased and that in *B* diminished that is the hot system has got hotter and the cold colder & yet no work has been done, only the intelligence of a very observant and neat fingered being has been employed.

If human ingenuity could 'apply tools to such portions of matter so as to deal with them separately', then it would be possible to manipulate the molecules individually, so that heat could be transferred from the cold to the hot system, in violation of the second law of thermodynamics. 'Only we can't, not being clever enough' (*LP*, **2**: 331–2).

It was not Maxwell's intention to raise the possibility of manipulating molecules, either by human ingenuity or by the agency of a 'finite being', so as to violate the second law of thermodynamics, While it would require the action of 'tools' or the 'finite being' to select individual molecules and produce an observable flow of heat, this process occurs spontaneously at the molecular level. Because of the statistical distribution of molecular velocities in a gas at equilibrium, there will be spontaneous fluctuations of molecules, fluctuations that take heat from a cold body to a hotter one. The purpose of the argument was to illustrate the conceptual status of the second law of thermodynamics. As he later noted to Tait, his purpose in invoking the thought-experiment had been 'To show that the 2nd law of Thermodynamics has only a statistical certainty'.[26]

The insight that Maxwell generates from this argument, which was consequent on the statistical distribution of the molecular velocities of a gas as postulated by his kinetic theory of gases, is that the second law of thermodynamics is a statistical law; it cannot be applied to the fluctuations of individual molecules. He explains this implication in a review of the second edition of Tait's *Thermodynamics*, in 1878. He notes that 'the second law of thermodynamics is continually being violated, and that to a considerable extent, in any sufficiently small group of molecules belonging to a real body'. This is because, according to Maxwell's statistical molecular theory, there are spontaneous fluctuations of molecules, which take heat from a cold to a hot body. But as the number of molecules in the group increases, deviations from the mean become less significant. Ultimately, the probability of a variation being measurable, producing an observable flow of heat from a cold system to a hot one and hence the violation of the second law of thermodynamics, 'may be regarded as practically an impossibility' (*SP*, **2**: 670).[27]

The name 'demon' for Maxwell's 'finite being' was coined by William Thomson, and first used by him publicly in 1874.[28] The name 'demon' did not receive Maxwell's approbation. In an undated note to Tait, 'Concerning Demons', he declares 'Call him no more a demon but a valve', and noting: 'Who gave them this name? Thomson'.[29] He was not concerned to establish the possibility of the existence and agency of the 'finite being', which was invoked in a thought-experiment, but to demonstrate the implications of his theory of the statistical distribution of molecular velocities in a gas at equilibrium. Thus the 'finite being' could be represented as a 'valve'; or, as he expressed it to J. W. Strutt (Lord Rayleigh) in December 1870, as a 'self-acting' device or 'guiding agent' (*LP*, **2**: 583). The purpose of the argument, as he explains in the note 'Concerning Demons', is 'To show that the 2nd law of Thermodynamics has only a statistical certainty'.

When Maxwell presents his argument in print in his *Theory of Heat* (1871), he does so to illustrate a 'Limitation of the second law of thermodynamics' (see Chapter VI.2). Earlier in the book, he adapts the argument to demarcate between the laws of mechanics, including the principle of the conservation of energy, and the consequences of the second law of thermodynamics.

> If heat in a body consists in a motion of its parts, and if we were able to distinguish these parts, and to guide and control their motions by any kind of mechanism, then by arranging our apparatus so as to lay hold of every moving part of the body, we could, by a suitable train of mechanism, transfer the energy of the moving parts of the heated body to any other body in the form of ordinary motion. The heated body would thus be rendered perfectly cold, and all its thermal energy would be converted into the visible motion of some other body.

But the second law of thermodynamics asserts that this process, which is 'consistent with the first law', the principle of conservation of energy, and the laws of dynamics, cannot occur. No 'train of mechanism' or 'any other method yet discovered' could perform this operation. The second law of thermodynamics is not therefore a dynamical law.[30]

He makes this point explicit in his discussion of the 'Limitation of the second law of thermodynamics'. This law could be abrogated by 'a being whose faculties are so sharpened that he can follow every molecule in its course'. The second law of thermodynamics applies to molecules in the mass, where deviations from the mean are insignificant: 'we are compelled to adopt what I have described as the statistical method of calculation, and to abandon the strict dynamical method' (see Chapter VI.2).[31] The second law of thermodynamics is a statistical law.

It was for this reason that Maxwell dismisses as illusory Clausius' attempt to reduce the second law of thermodynamics to a theory of molecular configuration. The concepts that Clausius introduced to provide such an interpretation – disgregation, ergal and virial – were subjected to jocular barbs (*LP*, **2**: 609, 709, 710, 946).[32] In his review of the second edition of Tait's *Thermodynamics* in 1878 he proposes that thermodynamics should be defined as 'the investigation of the dynamical and thermal properties of bodies, deduced entirely from what are called the First and Second laws of Thermodynamics, without any hypotheses as to the molecular constitution of bodies' (*SP*, **2**: 664–5).

Maxwell rejected current attempts, notably by Clausius and Boltzmann, to provide a general mechanical interpretation of the second law of thermodynamics, aiming to reduce the law to a theorem in dynamics. In December 1873 he commented to Tait on this reductionist project, with some asperity but also amusement.

> But it is rare sport to see those learned Germans contending for the priority of the discovery that the 2nd law of ΘΔcs is the Hamiltonsche Princip, when all the while they *assume* that the temperature of a body is but another name for the vis viva [kinetic energy] of one of its molecules, a thing which was suggested by the labours of Gay Lussac Dulong &c but first deduced from dynamical statistical considerations by d*p*/d*t* [Maxwell himself[33]]. The Hamiltonsche Princip, the while soars along in a region unvexed by statistical considerations while the German Icari flap their waxen wings in nephelococcygia [cloud- cuckoo-land] amid those cloudy forms which the ignorance and finitude of human science have invested with the incommunicable attributes of the invisible Queen of heaven. (*LP*, **2**: 947)

Clausius and Boltzmann were doomed to fall like Icarus, victims of the limitations of their science. Their proposed reduction of the second law of thermodynamics to a theorem in dynamics – 'as if any pure dynamical statement would submit to such an indignity', he remarked to Tait in October 1876[34] – was illusory, resting as it did on a fundamental misconception of the status of the law.[35] In his review of the second edition of Tait's *Thermodynamics* in 1878, he notes that the 'truth of the second law', as a statistical theorem, was 'of the nature of a strong probability . . . not an absolute certainty' like dynamical laws (*SP*, **2**: 671).

Maxwell broadens the argument, so as to highlight the distinction between reversible mechanical laws and the essential irreversibility of natural processes as described by the second law of thermodynamics. His first discussion of perfect reversibility comes in a letter to Thomson in November 1857, in response to Thomson's current speculations on hydrodynamics and his reference to Maxwell's paper on the curved trajectory of a strip of paper falling through the air (see Chapter V.2). Referring to their prior discussion, he notes that

> In May you thought that these effects would take place in an incompressible fluid without friction, and now you think that opinion a delusion, because if all motions at any instant were reversed all would go back as it came. Now I cannot see why, if you could gather up all the scattered motions in the fluid, and reverse them *accurately*, the strip should not fly up again. All that you need is to catch all the eddies, and reverse them not approximately, but accurately.
> (*LP*, **1**: 561–2)

These comments on the reversibility of the laws of dynamics were written before his engagement with the principles of thermodynamics. But in April 1868, writing to the Oxford scholar Mark Pattison about current speculations on the origins and ends of the universe (see Chapter IX.2), he explains that the time-reversal allowed by the reversible laws of dynamics is not compatible with the essential irreversibility of natural processes. In a universe regulated by the laws of dynamics alone,

> if every motion great & small were accurately reversed, and the world left to itself again, everything would happen backwards the fresh water would collect out of the sea and run up the rivers and finally fly up to the clouds in drops which would extract heat from the air and evaporate and afterwards in condensing would shoot out rays of light to the sun and so on. Of course all living things would regrede from the grave to the cradle and we should have a memory of the future but not of the past.

Of course this is not the world as we experience it, subject to the arrow of time.

> The reason why we do not expect anything of this kind to take place at any time is our experience of irreversible processes, all of one kind, and this leads to the doctrine of a beginning & an end instead of cyclical progression for ever. (*LP*, **2**: 360–1)

Maxwell's expression of the implications of time-reversal may perhaps echo the myth of the reverse motion of the universe expounded in Plato's *Politicus*,[36] but the implications for science are clear: natural processes are irreversible, and the theorems of dynamics do not describe the irreversibility of natural laws.

In a subsequent letter to Pattison, also dating from April 1868, Maxwell explains irreversible processes by using an analogy which directly illustrates the principles of his statistical theory of gases.

> Now in a nation you can pick out the rich people as such, but in a gas you cannot pick out the swift molecules either by mechanical or chemical means. As a simpler instance of an irreversible operation which (I think) depends on the same principle suppose so many black balls put at the bottom of a box and so many white above them. Then let them be jumbled together. If there is no physical difference between the white and black balls, it is exceedingly improbable that any amount of shaking will bring all the black balls to the bottom and all the white to the top again, so that the operation of mixing is irreversible unless

> either the black balls are heavier than the white or a person who knows white from black picks them and sorts them.
>
> Thus if you put a drop of water into a vessel of water no chemist can take out that identical drop again, though he could take out a drop of any other liquid. (*LP*, **2**: 366–7)

The irreversibility of these processes has only a statistical certainty; it is 'of the nature of a strong probability' but not 'an absolute certainty' (*SP*, **2**: 671). Unless there was a means of selecting and sorting the black and white balls, their jumbling would be effectively 'irreversible'. Further jumbling would be an 'exceedingly improbable' way to reverse the process and sort the balls back into their separate groups. By analogy, unless there was some 'mechanical or chemical means' – or, of course, the sorting agency of the 'finite being' – to select gas molecules, the mixing of fast and slow molecules would be 'irreversible', for the spontaneous separation of the swift molecules from the slow ones, to a degree that would have measurable effects, would be 'exceedingly improbable'. This letter to Mark Pattison was written after his letter to Tait (of December 1867) on the 'demon' paradox; while he does not mention the 'finite being', his argument is directed to establish the same point: that the irreversibility of the second law of thermodynamics has a statistical certainty.

He subsequently conjoins his arguments on the perfect reversibility of the motions of particles, the irreversibility of natural processes, and the statistical interpretation of the second law of thermodynamics. The discussion occurs in his letter to John William Strutt of December 1870.

> If this world is a purely dynamical system and if you accurately reverse the motion of every particle of it at the same instant then all things will happen backwards till the beginning of things the rain drops will collect themselves from the ground and fly up to the clouds &c &c and men will see all their friends passing from the grave to the cradle till we ourselves become the reverse of born, whatever that is. . . . The possibility of executing this experiment is doubtful but I do not think that it requires such a feat to upset the 2nd law of Thermodynamics. (*LP*, **2**: 582)

As Thomson had formulated it, the second law of thermodynamics describes the direction of the irreversible flow of heat in the universe, from hot to cold bodies. To illustrate the abrogation of this law Maxwell again introduces his 'finite being', now also described as 'a doorkeeper, very intelligent and exceedingly quick', as a 'mere guiding agent (like a pointsman on a railway with perfectly acting switches . . .)', and as a 'self-acting' device. This 'doorkeeper' could select molecules moving with different velocities so as to reverse the direction of heat flow, and so abrogate the second law of thermodynamics. The thought-experiment illustrates both the statistical nature of the second law of thermodynamics and its irreversibility. He draws the

> *Moral.* The 2nd law of Thermodynamics has the same degree of truth as the statement that if you throw a tumblerful of water into the sea you cannot get the same tumblerful out again. (*LP*, **2**: 582–3)

If it were possible to identify and pick out the molecules in the glass of water, then this process would be reversible. But as he had explained to Pattison, without such a method of selection the process is effectively 'irreversible'; its reversal would be 'exceedingly improbable' (*LP*, **2**: 367). The irreversibility of this process rests on the regularity of statistics, and provides an analogy for the irreversibility of the second law of thermodynamics. In a sufficiently small group of molecules, the spontaneous fluctuations among the molecules could reverse the direction of the flow of heat. But the probability of observing a measurable violation of the second law of thermodynamics, as the number of molecules in the group increases and deviations from the mean become insignificant, would be 'practically an impossibility' (*SP*, **2**: 670). The irreversibility of direction of the flow of heat, as described by the second law of thermodynamics, has a statistical certainty.

VII Geometry and physics

VII.1 Vectors: the geometry of field theory

In a letter to Peter Guthrie Tait in December 1867, Maxwell inquired of his friend: 'Is your book on Quaternions out yet?' (*LP*, **2**: 332–3). Tait's book, *An Elementary Treatise on Quaternions* (1867), was the product of his ten-year engagement with William Rowan Hamilton's calculus of quaternions, a form of vector analysis. Tait had been drawn to Hamilton's ideas because he perceived the utility of the method in mathematical physics. He drew Maxwell's attention to the application of Hamilton's operator ∇ in mathematical physics: 'If you read the last 20 or 30 pages of my book I think you will see that 4^{ions} are worth getting up, for there it is shown that they go into that ∇ business like greased lightning'.[1]

But it was not until the autumn of 1870, when he began to revise the manuscript of his *Treatise on Electricity and Magnetism*, published in 1873, that Maxwell's interest in the subject took fire. He became intrigued by the calculus of quaternions as a new mathematical method, a 'doctrine of Vectors', spatially directed quantities appropriate to the representation of physical concepts and relations. Noting that 'in electromagnetism we have a large number of different vector quantities', he perceived that electromagnetic quantities could be expressed as the 'Vector Functions of the Electromagnetic Field'. He sought to remould the mathematical argument of his book, and told Tait that he wished to use vector symbolism 'to make statements in electromagnetism', requesting Tait's advice as he did not wish to 'expose either myself to the contempt of the initiated, or Quaternions to the scorn of the profane' (*LP*, **2**: 569, 572, 951).

Hamilton had developed his calculus of quaternions from his work on algebra. He was concerned with the concept of number, attempting to give an algebraic definition of complex numbers and seeking to base algebra on the ordinal character of numbers. In his 'Essay on algebra as the science of pure time' (1835), he declared his aim: to establish algebra as a 'SCIENCE, *in some sense analogous to Geometry*'. The argument is Kantian in inspiration; appealing to Kant's doctrine of space and time as forms of intuition, Hamilton claims that 'the intuition of TIME' provides the basis of a 'SCIENCE of algebra'. He maintains that the 'notion of intuition of ORDER IN TIME is not less but more deep-seated in the human mind than the notion or intuition of ORDER IN SPACE'. Kant's philosophical language provides the basis for the analogy between geometry and algebra. Just as the intuition of space establishes the basis of geometry, 'the Science of Pure Time . . . is co-extensive and identical with Algebra, so far as Algebra itself is a Science'.[2]

Hamilton aimed to extend the complex number system to three dimensions; this led, in 1843, to his invention of quaternions, hypercomplex numbers with one real and three (imaginary) complex parts. He interpreted the three imaginary numbers as 'vectors' directed along three mutually perpendicular lines in space. The fourth and real part of the quaternion corresponded to a line in space of only one dimension: this was the 'scalar' part of the quaternion. A quaternion was therefore conceived as the sum of its own vector and scalar parts. The quaternion calculus, he wrote,

> selects *no one direction in space as eminent* above another, but treats them as all equally related to that *extra-spatial* or simply SCALAR direction . . . .[3]

All directions in space are related to the 'scalar' in the same way. The 'scalar' is a line but not a line in three-dimensional space, and is non-directional. Quaternions provide a form of analytical geometry which requires no prior selection of coordinates.

Hamilton devoted considerable attention to the interpretation of the meaning of quaternions, seeking to conceptualise the addition of quantities (vector and scalar) of different dimensions. In his review of Kelland and Tait's *Introduction to Quaternions* (1873) Maxwell drily observes that

> Sir W. R. Hamilton, when treating of the elements of the subject, was apt to become so fascinated by the metaphysical aspects of the method, that the mind of his disciple became impressed with the profundity, rather than the simplicity of his doctrines. (*LP*, **2**: 952)

Here Maxwell was very likely making allusion to Hamilton's elaborate geometrical illustrations of the algebraic properties of quaternions in the 'Preface' to his *Lectures on Quaternions* (1853), symbolic arguments which he had described as 'partly geometrical, but partly also metaphysical (or *à priori*)'. In the 'Preface' to his *Lectures* Hamilton repeated his notion of 'Algebra as the SCIENCE OF PURE TIME', and referred to Kant in support of the contention that 'it should be *possible* to construct, *à priori*, a Science of Time, as well as a Science of Space', reinforcing the philosophical resonances of his work.[4]

Maxwell did not, however, wholly disparage Hamilton's metaphysics. Writing to Mark Pattison in April 1868, he observed that the 'Edinburgh & the Dublin Hamilton differ in their metaphysical power in the direct ratio of their physical knowledge (not the inverse as most people suppose)' (*LP*, **2**: 361), favourably contrasting the metaphysics of Hamilton the mathematician with that of his Edinburgh professor Sir William Hamilton (see Chapter II.3). But he was concerned to endorse the mathematical simplicity of Hamilton's ideas, to blunt dismissal of quaternions on grounds of seeming opacity. The 'doctrine of Vectors', as he describes the calculus of quaternions, had the virtue of conceptual clarity in its application to physics. As he made clear in his 1873 review of Kelland and Tait's book, he found quaternions valuable because

> the method . . . easily . . . arrives at those solutions which have been already expressed in ordinary mathematical language, and . . . brings within our reach other problems, which the ordinary methods have hitherto abstained from attacking.

In espousing quaternion methods he was not concerned with Hamilton's achievement in pure mathematics, in creating an algebra which does not obey the commutative law of multiplication, thereby demonstrating that algebras following alternative rules could be constructed. For Maxwell, quaternions provide a geometrical way of thinking.

> Now quaternions, or the doctrine of Vectors, is a mathematical method, but it is a method of thinking, and not, at least for the present generation, a method of saving thought . . . . It calls upon us at every step to form a mental image of the geometrical features represented by the symbols, so that in studying geometry by this method we have our minds engaged with geometrical ideas, and are not permitted to fancy ourselves geometers when we are only arithmeticians. (*LP*, **2**: 951–2)

The problems in which he was especially interested were those of mathematical physics, to which he refers in concluding his review. Vectors could provide a 'mental image', a geometrical representation of physical quantities. Geometry is seen as the keystone of intelligibility, an argument which echoes the traditional emphasis of Scottish mathematics and philosophy (see Chapter II.3); and his commitment to representation by means of 'geometrical ideas' evokes the perspective of his first paper on field theory, 'On Faraday's lines of force' (see Chapter IV.2).

Maxwell's interest in quaternions flowed from his concern, in revising his *Treatise on Electricity and Magnetism* for publication, to remould its mathematical argument to incorporate the physical geometry which he had espoused in his paper 'On Faraday's lines of force'. By embedding the geometry of vectors within the mathematical argument of the *Treatise*, he transformed and enlarged his geometrical representation of physical quantities. In the mathematical 'Preliminary' to the book he claims that the quaternion method provides a direct representation of electrical quantities congruent with their physical meaning. The method is a means of expressing geometrical and physical quantities which is 'more primitive and more natural' than the method of Cartesian coordinates (*Treatise*, **1**: 8–9 (§10)). The scope of this newly aroused interest in quaternions is made clear in his letters to Tait in late 1870 and early 1871; in a manuscript draft on the 'application of the ideas of the calculus of quaternions to electromagnetic phenomena' of November 1870 (*LP*, **2**: 568–79, 590–7, 609); and in his paper 'On the mathematical classification of physical quantities' (*SP*, **2**: 257–66), read to the London Mathematical Society in March 1871.

In his first letter to Tait on the subject, in November 1870, he wrote out the quaternion expression for Hamilton's operator, in the form as given by Tait:[5]

$$\nabla = i\frac{d}{dx} + j\frac{d}{dy} + k\frac{d}{dz}$$

where *i*, *j*, *k* are unit vectors parallel to the axes *x*, *y*, *z*. He inquired about the name of the operator, written as an inverted delta: 'What do you call this? Atled?' (*LP*, **2**: 568). Hamilton had first stated this expression for his 'characteristic of operation' in 1846, and in his *Lectures on Quaternions* had noted that this function and its square ∇^2 would be '*extensively useful in the mathematical study of nature*', making reference to its occurrence in 'the modern researches in ANALYTICAL PHYSICS, respecting attraction, heat, electricity, magnetism, &c',[6] presumably having in mind the theorems of Laplace and Poisson in potential theory.

As Tait explained in his paper 'On Green's and other allied theorems' (1870), it was these expressions and their prospective application in mathematical physics – which forms the subject of his paper – that had first aroused his interest in quaternions. Maxwell admired Tait's paper, which may well have provoked his serious interest in quaternions, and which in November 1870 he eulogised as 'really great' (*LP*, **2**: 581). Tait begins by stating the expression for Hamilton's operator ∇, citing Hamilton's remark on 'its promise of usefulness in physical applications'. In illustration he expresses Green's and Stokes' theorems in quaternion form; and points to the 'simplicity and expressiveness of quaternions' in establishing the 'mutual relationship' of the properties of the 'analytical and physical magnitudes which satisfy what is usually known as Laplace's equation', terming ∇^2 'Laplace's operator'.[7]

Because electromagnetic quantities are vectors, Maxwell was interested in the calculus of quaternions as a method in mathematical physics. He presents the calculus of quaternions as the 'doctrine of Vectors'. His interest in quaternions was specific: 'the value of Hamiltons idea of a Vector is unspeakable', he told Tait, and so 'I want to leaven my book [the *Treatise*] with Hamiltonian ideas without casting the operations into a Hamiltonian form'. Thus, in his draft of November 1870 on the application of quaternions to electromagnetism, he lists the 'Vector Functions of the Electromagnetic Field' (*LP*, **2**: 572, 577); he is concerned with the physical meaning of vectors rather than the mathematical ramifications of quaternion methods.

He drew upon this draft in writing the mathematical 'Preliminary' to the *Treatise.* The advantage of vectors over Cartesian coordinates, he explains, is that the vectorial methods enable the investigator

> to fix the mind at once on a point of space instead of its three coordinates, and on the magnitude and direction of a force instead of its three components.

He maintains that the traditional component expressions were 'really the most useful for purposes of calculation', and hence it would be inappropriate to deploy the full array of quaternion algebra in a treatise on mathematical physics. But he urges that the 'ideas, as distinguished from the operations and methods of Quaternions, will be of great use . . . in electrodynamics' (*Treatise,* **1**: 9 (§10)). He espouses quaternions as a mode of thinking, as inherently geometrical, a point strongly expressed in his draft:

> The calculus of quaternions is not therefore like the Cartesian geometry a method of applying the science of number to the investigation of space, it is a calculus founded on the independent investigation of space and in which several of the rules relating to operations with numbers are not applicable. (*LP*, **2**: 571)

It is for this reason, he explains in the *Treatise*, that the spatial, geometrical language of vectors could denote the physical relations of electromagnetism 'by a few words'. The vectorial, geometrical method is therefore 'more primitive and more natural' a 'mode of contemplating geometrical and physical quantities' than the traditional method using Cartesian coordinates and component notation (*Treatise*, **1**: 9 (§10)).

Maxwell is especially concerned with quaternion nomenclature. The Hamiltonian operator ∇ was named 'Nabla' (an Assyrian harp), and Tait became (in jocular reference to the 'Book of Psalms') the 'Chief Musician upon Nabla' (*LP*, **2**: 577).[8] But his especial concern, as he explained to Tait, was to settle on names for the result of Hamilton's operator on scalar and vector functions. The effect of ∇ on a vector function was especially significant, giving quaternion expressions for the solenoidal and lamellar distributions of magnetism (see Chapter IV). As he explains in his paper to the London Mathematical Society:

> If σ represents a vector function, $\nabla\sigma$ may contain both a scalar and a vector part, which may be written $S\nabla\sigma$ and $V\nabla\sigma$. I propose to call the scalar part the *Convergence* of σ . . . a very good name for the effect of that vector function in carrying its subject inwards towards a point. But $\nabla\sigma$ has, in general, also a vector portion, and I propose, but with some diffidence to call this vector the *Curl* or *Version* of the original vector function. It represents the direction and magnitude of the rotation of the subject matter carried by the vector σ. (*SP*, **2**: 265)

The expressions $S\nabla\sigma = 0$ and $V\nabla\sigma = 0$ denote the solenoidal and lamellar distributions of magnetism, respectively. [These quaternion expressions translate directly into their equivalents in vector analysis.]

The terms 'convergence' and 'curl' are geometrical, and were chosen to indicate direction. The term 'curl' caused him some deliberation:

> It might be called rotation, version, twist or twirl but all these names have motion implied in them so that I prefer the word curl.

The geometrical 'curl' was to be preferred to the dynamical 'twirl' (*LP*, **2**: 569, 574). In 'On physical lines of force' (1861–2), he had posited a relation between the electric current and the magnetic force (*SP*, **1**: 462, 502), which embodies the mechanical imagery employed in the paper: the translation of the 'idle wheel' particles in his ether model and the rotation of molecular vortices (see Chapter V.2). In the *Treatise*, this relation is expressed in quaternion form between vectors **C** and **H** (electric current and magnetic force) as $4\pi\mathbf{C} = V\nabla\mathbf{H}$, the equation of 'electric currents', equation (E) of Maxwell's 'General equations of the electromagnetic field' (*Treatise*, **2**: 238 (§619)). He uses quaternion notation to express the rotational character of magnetism by the expression $V\nabla\mathbf{H}$, the 'curl' of the vector function **H**, the magnetic force.

The quaternion symbolism denotes the relation between physical quantities and directions in space, and does not embody a dynamical representation. In the dynamical theory of the *Treatise* (see Chapter VI.1), the physical geometry of quaternions was to be supplemented by dynamical explanation. In the 'Preliminary' to the *Treatise*, where he discusses the application of vectorial methods, he places special emphasis on the 'distinction between longitudinal and rotational properties', which, he observes, is 'very important in a physical point of view'. While the electric current is a longitudinal phenomenon, 'the effect of magnetism in rotating the plane of polarized light distinctly shews that magnetism is a rotational phenomenon' (*Treatise*, **1**: 12–13 (§15)). In a postcard written to William Thomson, early in 1873, he observes that

> It is very remarkable that in spite of the *curl* in the electromagnetic equations of all kinds Faradays twist of polarized light will not come out without what the schoolmen call local motion. (*LP*, **2**: 784)

Quaternion symbolism expresses the physical geometry of electromagnetism, not the dynamical properties of the electromagnetic field. In the *Treatise* he continues to affirm that the Faraday effect was to be explained by the 'rotation . . . of very small portions of matter' which depend on 'some kind of mechanism connecting them' (*Treatise*, **2**: 416 (§831)). Quaternion symbolism, the 'doctrine of Vectors', expresses the geometrical basis of this dynamical theory of the electromagnetic field.

In the mathematical 'Preliminary' to the *Treatise* Maxwell uses the concept of vectors to reformulate the distinction between 'intensities' and 'quantities' that he had drawn in his paper 'On Faraday's lines of force' (see Chapter IV.2). There he had distinguished between electrical and magnetic quantities acting along lines (defined as 'intensities') and those acting across surfaces (defined as 'quantities'). He reformulates this distinction in the *Treatise* in terms of a classification of vectors.

> Physical vector quantities may be divided into two classes, in one of which the quantity is defined with reference to a line, while in the other the quantity is defined with reference to an area.

A force acting on a body is defined with reference to a line, while the flux of heat or electricity is defined with reference to an area. In the science of electricity and magnetism, electromotive and magnetic forces are defined with reference to lines; these he terms 'Forces'. Electric displacement, electric current and magnetic induction are defined with reference to areas; these he terms 'Fluxes'.[9] 'Each of these forces', he explains, 'may be considered as producing, or tending to produce, its corresponding flux' (*Treatise*, **1**: 10–11 (§12)).

He goes on to discuss two important mathematical operations defined in relations to these two classes of vectors. The first is the 'Line-integral of a force': the integral along a line of the resolved part of the 'force'. The second is the 'Surface-integral of the flux': the integral over a surface of the 'flux' through every element of the surface. The description of vectors as 'forces' and 'fluxes', acting along lines or across areas, opened the way to the introduction of line- and surface-integrals, and to a further conceptual enlargement of the role of vector quantities in the mathematical theory of electromagnetism: the explication of Stokes' theorem, which establishes a relation between the surface-integral of a flux taken over a surface and the line-integral of a force taken round the boundary of the surface. The electromagnetic law which he had formulated in 'On Faraday's lines of force', which expresses the relation between the flux of magnetic induction (now represented by the vector **B**) through a surface and the electro-tonic intensity (now represented by the 'vector potential' **A**) round its boundary, is, in the *Treatise*, expressed in terms of vectors and Stokes' theorem.

This theorem had been stated by Thomson and Tait in their *Treatise on Natural Philosophy* (1867),[10] though Maxwell remembered that Stokes had set the theorem in his Smith's Prize examination of 1854 (see Chapter IV.2). 'Did not you set the theorem about the surface integral . . . over a surface bounded by . . . [a] curve. . . being equal to [its corresponding line integral?]', he inquired of Stokes in January 1871. He added that 'I have had some difficulty in tracing the history of this theorem. Can you tell me anything about it [?]' (*LP*, **2**: 589). Stokes' memory may have been rusty after twenty years – the theorem had been first stated by Thomson in a letter to him in July 1850[11] – and he was apparently unable to answer Maxwell's query, for Maxwell raised the matter with Tait the following April. Tait had given the theorem in quaternion form in his paper 'On Green's and other allied theorems', and in response to Maxwell's remark that the theorem 'ascends (at least) to Stokes Smiths Prize paper 1854 and it was then not altogether new to yours truly' (*LP*, **2**: 634), he replied that he had thought the theorem to be Thomson's and first published in their *Natural Philosophy*.[12]

On proving the theorem in the 'Preliminary' to the *Treatise*, Maxwell attributes it to Stokes and ascribes its proof to Thomson and Tait (*Treatise*, **1**: 27n (§24)). He states the theorem in the following form: 'the surface-integral of **B** taken over the surface *S* is equal to the line-integral of **A** taken round the curve *s*', where *s* is a closed curve and *S* a continuous finite surface bounded entirely by *s* (*Treatise*, **1**: 25 (§24)). In stating the theorem in terms of relations between vectors **A** (a 'force') and **B** (a 'flux), which have as yet undefined physical meaning, he clearly has an eye for the application of the theorem to electromagnetism. On expounding its physical significance later in the book, he notes that the 'vector **A** represents . . . the time-integral of the electromotive force which a particle . . . would experience if the primary current were suddenly stopped', while 'we must regard the vector **B** . . . as representing . . . the magnetic induction'. He terms **A** the 'vector-potential of magnetic induction', and it corresponds to the 'electro-tonic intensity' introduced in 'On Faraday's lines of force' (*Treatise*, **2**: 28, 214–16(§§405, 590, 592)).

He points out that the electromagnetic force and magnetic induction in a circuit depend on the variation in the number of magnetic lines of force which pass through the circuit, and that

> the number of these lines is expressed mathematically by the surface-integral of the magnetic induction through any surface bounded by the circuit.

Thus, when applied to electrical circuits and electromagnetic induction, the vectors **A** and **B** have their electromagnetic meaning, as the 'vector potential' and 'magnetic induction', and Stokes' theorem may be stated:

> we may express the line-integral of **A** round any circuit in the form of the surface-integral of **B** over a surface bounded by the circuit.
>
> (*Treatise*, **2**: 215–16 (§§591–2))

The electromagnetic relation between magnetic induction through a surface and the electro-tonic intensity round its boundary, first stated in 'On Faraday's lines of force', is now formally expressed in terms of vectors ('forces' and 'fluxes') and Stokes' theorem. This is not merely an analytic reformulation of an electromagnetic 'law' originally expressed verbally: the meaning and rationale of this electromagnetic relation have been transformed in the process of conceptual enlargement. Maxwell's expression of the theory of electromagnetism in terms of vectors, 'forces' and 'fluxes', and line- and surface-integrals, grounded the theory on a framework of purely geometrical relations. He thus transforms the physical geometry of 'On Faraday's lines of force' by appeal to a 'mathematical method specially adapted to the expression of such geometrical relations – the *Quaternions* of Hamilton' (*Treatise*, **2**: 159 (§522)).

VII.2 'Geometry of position': topology and projective geometry

Writing to Tait in late May 1871 Maxwell remarked that

> Carnots Geometry of Position is like Chasles Superior ditto. Mine is that which Gauss calls Geometria Situs as opposed to G. Magnitudinis.
>
> (*LP*, **2**: 646–7)

The distinction to which he alludes is between the projective geometry of Lazare Carnot (in his *Géométrie de Position* of 1803) and Michel Chasles (in his *Traité de Géométrie Supérieure* of 1852), and the new branch of mathematics which had been termed 'topology' by Johann Benedict Listing in his 'Vorstudien zur Topologie' of 1847.[13] Maxwell's term 'geometry of position' was, as he indicated, drawn from Gauss' usage. In a manuscript note of 1833, probably occasioned by reading Leibniz's remarks on the need to formulate geometric algorithms to express geometrical location (*situs*), Gauss had discussed the geometrical connection between two closed curves as an example of *Geometria Situs.* Gauss' note was included in the fifth volume of his collected papers, published in 1867,[14] and his argument shaped Maxwell's discussion of the topology of 'knotted curves' in a letter to Tait in December 1867 (*LP*, **2**: 325–7), and subsequently in the *Treatise.*

In declaring his adherence to the 'geometry of position' in the sense of topology, Maxwell was making specific allusion to issues raised in his correspondence with Tait earlier that month. They had discussed the convention specifying the relation between linear and rotational motions in space. In seeking to define the convention to be used in the *Treatise*, an issue crucial to understanding the relation between lines of force and electrical circuits, Maxwell had drawn on Listing's topology (*LP*, **2**: 641). In the *Treatise* he enlarges the physical geometry of his paper 'On Faraday's lines of force' to include the topological treatment of lines and surfaces and the expression of their electromagnetic analogues.

But the 'geometry of position', understood in the sense intended by Lazare Carnot, as projective geometry, also plays an important part in Maxwell's mathematical physics. He was familiar with the projective geometry of J. V. Poncelet and Michel Chasles, and he welcomed this revival of projective geometry because of the primacy he accorded to geometrical over analytic modes of thinking. Writing in 1873, he explains that

> [the] study of corresponding elements in two figures has led to the establishment of a Geometry of Position by which results are obtained by pure reasoning without calculation the verification of which by the Cartesian analysis would fill many pages with symbols. (*LP*, **2**: 899)

He uses concepts drawn from projective geometry as a method of geometrical analogy – a method first discussed in 'On Faraday's lines of force' (see Chapter IV.2) – to be applied to the graphical analysis of frames, geometrical optics, electrical circuits, and the kinetic theory of gases.

The origins of topology lie in Leibniz's interest in the formulation of geometrical algorithms to express geometrical location; in Euler's work on the classification of polyhedra in terms of a relation between the number of edges, faces and vertices of closed convex polyhedra; and, most importantly for Maxwell, in Bernhard Riemann's recent discussion of continuity in geometry, where he had attempted to classify surfaces and their boundaries by their topological properties. Maxwell very likely became aware of Riemann's work through Helmholtz's paper on vortex motion, published in 1858 and in English translation by Tait in 1867.[15] Helmholtz had made use of Riemann's classification of surfaces by their topological connectivity, an argument developed in the course of his work on complex function theory.[16] William Thomson had drawn on Helmholtz's paper, and used Riemann's terminology, in his paper 'On vortex motion' which was read to the Royal Society of Edinburgh in April 1867.[17]

Writing to Tait in November 1867, Maxwell referred to Helmholtz's and Thomson's papers, and discussed the knotting of vortices (*LP*, **2**: 321–2). He developed the argument in a further letter to Tait written the following month: 'I have amused myself with knotted curves for a day or two'. His ideas were shaped by reading Gauss' note on *Geometria Situs*, expressing the integral round two interlinked closed curves. He drew various forms in which the curves could be knotted, noting the analogy to problems in electromagnetism (*LP*, **2**: 325–7). He discusses Gauss' comments in the *Treatise*, drawing attention to their bearing on electromagnetism:

> It was the discovery by Gauss of this very integral, expressing the work done on a magnetic pole while describing a closed curve in presence of a closed electric current, and indicating the geometrical connexion between the two closed curves, that led him to lament the small progress

made in the Geometry of Position since the time of Leibnitz, Euler and Vandermonde. We have now, however, some progress to report, chiefly due to Riemann, Helmholtz and Listing.

(*Treatise* **2**: 41(§421))

He continued to explore the relation between Helmholtz's theorems of vortex motion and the theory of electromagnetism (*LP*, **2**: 398–407), and discussed the topology of curves and surfaces (*LP*, **2**: 433–42, 444, 449), noting that the relations of bounding surfaces and closed curves were 'important in the theory of vortices and in Electromagnetism' (*LP*, **2**: 441). This interest culminated in a draft of December 1868 on the application of Listing's topology to the relations between geometrical figures (*LP*, **2**: 466–9). He presented an account of Listing's memoir 'Der Census räumlicher Complexe' [Survey of spatial complexes] (1861)[18] to the London Mathematical Society in February 1869, noting Gauss' theorem on the topology of knotted curves (*LP*, **2**: 470–1). He set a question on the subject for the Cambridge Mathematical Tripos,[19] and wrote a paper on topographical geometry in 1870, applying topological concepts to maps and contour lines, to the physical geography of 'hills and dales' (*LP*, **2**: 566–7; *SP*, **2**: 233–40). His drafts show him grappling with the problems of curves and surfaces, and adopting Listing's complex terminology for the classification of the geometrical relations between boundaries and surfaces.

These topological ideas are presented in some detail in the *Treatise*, where he deploys definitions of lines, boundaries and surfaces, emphasising their application to the construal of line- and surface-integrals. Discussion of the number of surfaces bounding finite regions of space; of the continuity of surfaces and their connectivity by closed curves; of the closure of finite surfaces by external surfaces; and definitions of the continuity of lines and surfaces, are set out in the mathematical 'Preliminary' to the *Treatise*, providing a geometrical basis for the interpretation of integral theorems and their application to the physics of electromagnetism (*Treatise*, **1**: 16–24 (§§18–22)). Vectors, integral theorems and topological concepts form the mathematical foundations of the *Treatise*, a framework of concepts expressing the geometrical and vectorial character of electrical and magnetic quantities.

Topological ideas enter into Maxwell's definition of the relation between lines of force and electric currents. As with his introduction of vectors, the issue arose during the revision and amplification of the mathematical argument of the *Treatise*. Once again he raised the matter with Tait, for he had noticed a lack of consistency between Hamilton's *Lectures on Quaternions* and Thomson and Tait's *Treatise on Natural Philosophy*, in defining the convention relating motion along an axis and the rotation around it.[20] The problem surfaced in May 1871, at around the time he signed a contract with the Clarendon Press for the publication of the *Treatise* (*LP*, **2**: 636). Addressing Tait, he expresses his confusion in the jocular and allusive style customary in their correspondence: 'I am desolated! I am like the Ninevites! Which is my right hand? Am I perverted? a mere man in a mirror, walking in a vain show?' (*LP*, **2**: 637). Following discussion at the London Mathematical Society he soon clarified the issue, and wrote out for Tait's approval a draft of §23 of the *Treatise.*

> In this treatise, the motions of translation along any axis and of rotation round that axis will be assumed to be of the same sign when they are related to each other in the same way as the motions of translation and rotation of a right handed screw. . . . This is the right handed definition of directions and is adopted in this treatise. . . .

While this was the system adopted by Thomson and Tait, the opposite system had been adopted by Hamilton, and Maxwell explains that 'If we confound the one with the other, every figure will become *perverted* (a phrase of L[isting] denoting an effect similar to that of reflexion in a mirror)' (*LP*, **2**: 644).

Listing had introduced the term *Perversion* in his 'Vorstudien zur Topologie', to denote the operation of passing between an object and its reflected image; in using this term to describe the relation between the right-handed and the left-handed system of rotation, Maxwell emphasised the topological basis of his convention for the relation between translation along an axis and rotation about that axis. He describes the system he adopts, that of the right-handed screw, as that of the 'tendril of the vine'; the opposite, left-handed system he describes as being that of the 'tendril of the hop' (*LP*, **2**: 641). This designation was suggested by W. H. Miller, the Cambridge author of *A Treatise on Crystallography* (1839), a work cited by Listing in his 'Vorstudien zur Topologie' for discussion of crystal symmetry. Listing himself had cited the direction of tendrils in botany in discussing the handedness of helices.[21]

Maxwell gives an account of this terminology in the *Treatise*, affirming that a 'common corkscrew may be used as a material symbol of the same relation', that of the right-handed screw or 'system of the vine' (*Treatise*, **1**: 24n (§23)). To Tait he wrote

> Lastly I thank you and praise you for turning me from the system of the hop to that of the vine. I have perverted the whole of electromagnetics to suit. (*LP*, **2**: 645)

Thus the corkscrew rule as the 'system of the vine' would specify the relation between the directions of an electric current flowing in a circuit, the magnetic field, and the electromagnetic force acting on the wire (*Treatise*, **2**: 143 (§498); *LP*, **2**: 679).

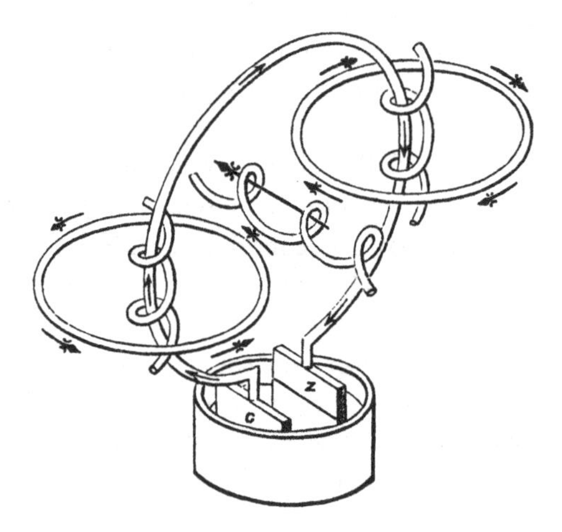

Fig. VII.1. The figure shows the relation between the electric current in a circuit generating a magnetic field, and the direction of the mechanical force acting on the conductor carrying the current. The electric current is longitudinal while magnetism is a rotational phenomenon, and the relation between the direction of the current and the lines of magnetic force is indicated by a right-handed screw. From *Treatise*, **2**: 143 (§498).

Maxwell's account of the Faraday magneto-optic effect (see Chapter V.2) in the *Treatise* provides an important illustration of his use of topological concepts. He points out that a plane-polarised ray of light can be represented by two circularly polarised rays, one right-handed, the other left-handed (as regards the observer).

> Any undulation, the motion of which at each point is circular, may be represented by a helix or screw . . . [and] the propagation of the undulation will be represented by the apparent longitudinal motion of the similarly situated parts of the thread of the screw.

A plane-polarised ray can thus be represented by a left-handed and a right-handed helix. This geometrical representation of the two circularly polarised rays shows that rays of the same wavelength 'are geometrically alike in all respects, except that one is the *perversion* of the other, like its image in a looking-glass'.

He maintains that the Faraday effect cannot be explained simply on the supposition that one of these rays has a shorter period of rotation than the other. The Faraday effect involves rotation of the direction of light:

> Hence greater forces are called into play when the helix is going round one way than when it is going round the other way. The forces, therefore, do not depend solely on the configuration of the ray, but also on the direction of the motion of its individual parts.
>
> (*Treatise*, **2**: 403–4 (§§813–15))

Because the propagation of light in a medium under the action of a magnetic force is 'affected by the relation of the direction of rotation of the light to the direction of the magnetic force', he concludes that 'some rotatory motion is going on'. He comes

> to the conclusion that in a medium under the action of magnetic force something belonging to the same mathematical class as an angular velocity, whose axis is in the direction of the magnetic force, forms a part of the phenomenon.

The rotation of light in the Faraday effect, conceived mathematically in terms of an angular velocity, requires dynamical explanation in terms of the motion of the medium.

> We must therefore conceive the rotation to be that of very small portions of the medium, each rotating on its own axis. This is the hypothesis of molecular vortices. (*Treatise*, **2**: 407–8 (§§821–2))

Maxwell's geometrical and dynamical arguments thus coalesce in his explanation of the Faraday effect. As he points out to Thomson early in 1873, 'Faradays twist of polarized light will not come out without . . . local motion' (*LP*, **2**: 784). The geometrical and dynamical arguments of the *Treatise* lead him, in his explanation of the Faraday magneto-optic effect, to expound the 'hypothesis of molecular vortices' first advanced in 'On physical lines of force' in 1861–2 (see Chapter V.2).

The physical geometry of field theory is, in the *Treatise*, explicated in terms of topology and the geometry of vectors. In the 1850s, the geometrical analogy between lines and surfaces, fluid flow and the electromagnetic field shaped the expression of his field theory as a physical geometry in 'On Faraday's lines of force' (see Chapter IV.2). Geometrical analogy continued to play a significant role in shaping Maxwell's physics, notably in his application of the methods of projective geometry.

He first drew upon these ideas in applying graphical analysis to the theory of frameworks. He lectured on the theory of engineering structures to his class at King's College London in the early 1860s (*LP*, **2**: 60), and went on to write a paper on the subject (*SP*, **1**: 514–25); and he subsequently extended his theory of reciprocal diagrams in statics (*LP*, **2**: 313–17; *SP*, **2**: 161–207). The work drew on projective geometry, which he terms the 'geometry of position' and 'modern geometry', and he makes especial use of the concept of geometrical correspondence between reciprocal figures. This method, he explains in 1873, on applying it to geometrical optics, is based on the 'principle of duality . . . the leading idea of modern geometry' (*LP*, **2**: 899, 935, 942).

Maxwell began to grasp the prospect of developing the conceptual range of projective geometry as a method of analogy and correspondence, focusing on the concept of reciprocity and the principle of duality, introduced by Poncelet and Chasles. In September 1867 he read a paper to the British Association for the Advancement of Science on the application of the theory of 'reciprocal polar figures' to the graphical analysis of frameworks (*LP*, **2**: 313–17). The same month he wrote to Thomson pointing out the analogy between reciprocal relations in elastic frameworks and in electric circuits. He suggests an analogy between the operation of pressure and its effect, and Green's reciprocity theorem between potential and charge (*LP*, **2**: 320).

In 1873 he outlined an extension of the method of 'corresponding elements in two figures' to the depiction of the velocities of molecules in a gas. He uses Hamilton's geometrical representation of the paths of particles (the hodograph), in which the successive velocities of a particle are depicted by a series of lines.[22] In this method, 'by studying the correspondence of these curves the force acting on the body and the whole circumstance of the motion may be ascertained'. In this way the kinetic theory of gases would be given a geometrical interpretation.

> We may regard this method as an example of one of the most powerful instruments of mathematical research – the simultaneous contemplation of two systems so related to each other that every element in the one has its corresponding element in the other.

This is the 'Geometry of Position', and is based on geometrical correspondence. He had employed it in his theory of stresses in frameworks, but it had wider application, in geometrical optics. He notes that

2 Topology and projective geometry

> In Geometrical Optics the study of the correspondence between the object and the image has been of almost equal service to the theory of optical instruments and to pure geometry. (*LP*, **2**: 899)

Maxwell had been interested in geometrical optics and in the theory of optical instruments since the 1850s, when he had expounded a theory of optical instruments based on the geometrical relations of light rays; geometrical optics is here divorced from the dioptrical properties of lenses (*SP*, **1**: 271–85). In September 1873 he developed this work, presenting a paper to the British Association on the application of the principle of duality, the correspondence between geometrical figures, to geometrical optics. He realised that geometrical optics could be placed within a broader mathematical framework. As he explained to J. W. Strutt (now Lord Rayleigh), 'I am getting more light on Geometrical Optics', having grasped that 'the geometry of the subject is the geometry of position' (*LP*, **2**: 942). In his paper he points out that every straight line or ray in the object is represented by a straight line or ray in the image, and that this is an example of the 'relations between pairs of figures' elaborated by the 'geometry of position':

> elementary optics might be made attractive to the mathematician by showing that the correlation between the object and the image is not only an example, but the fundamental type of that principle of duality which was the leading idea of modern geometry. (*LP*, **2**: 935)

Maxwell presents the principle of reciprocity or duality as a method of analogy, based on the correspondence and relation between geometrical figures. He first deploys the method in the graphical analysis of frames, and then illustrates its application in geometrical optics, in the study of electrical circuits, and the theory of gases. The appeal to geometrical representation and analogy, the use of geometry as a means of expressing physical relations, and the method of analogical reasoning, which have paramount importance in the argument of 'On Faraday's lines of force' (see Chapter IV.2), continued to be significant themes in his natural philosophy. Geometry and analogical reasoning are seen as the guide to the intelligibility of physical concepts (see Chapter II.3). As he expresses it in his essay on analogies, written for the Cambridge Apostles in February 1856, 'in a scientific point of view the *relation* is the most important thing to know' (*LP*, **1**: 382).

VIII Physical reality: ether and matter

VIII.1 Ether, field, and gravity

Maxwell's statement in the third part of his paper 'On physical lines of force' in 1862, that '*light consists in the transverse undulations of the same medium which is the cause of electric and magnetic phenomena*' (*SP*, **1**: 500), implied the unification of optics and electromagnetism in terms of a mechanical theory of an ether that had both optical and electromagnetic correlates. In 'A dynamical theory of the electromagnetic field' (1865), where he derives the wave propagation from the field equations, he still appeals to an ether as the dynamical basis of his unified theory of optics and electromagnetism. He now terms his theory the 'Electromagnetic Theory of Light':

> light and magnetism are affections of the same substance, and . . . light is an electromagnetic disturbance propagated through the field according to electromagnetic laws. (*SP*, **1**: 577, 580)

In the *Treatise on Electricity and Magnetism* (1873) he explains that the consilience of the evidence from optics and electromagnetism produced 'a conviction of the reality of the medium'. The supposition of an ether might be judged hypothetical, but its existence was rendered plausible by his demonstration that the velocity of propagation of electromagnetic waves was close to the velocity of light. He observes that

> if the study of two different branches of science has independently suggested the idea of a medium . . . of the same kind . . . the evidence for the physical existence of the medium will be considerably strengthened. (*Treatise*, **2**: 383 (§781))

Although Maxwell does not use Whewell's terminology, his argument echoes Whewell's key philosophical doctrine of the '*Consilience of Inductions*'. Whewell maintained that the grandest theories were produced when separate inductions from 'classes of facts altogether different have . . . *jumped together*'. Gravity and the wave theory of light, where disparate phenomena were subsumed under unifying explanatory structures, provided his main examples of theoretical convergence. Consilience led to unification and simplification, for 'there will be no need for new machinery in the hypothesis'. He claims that the 'Consiliences of our Inductions give rise to a constant Convergence of our

Theory towards Simplicity and Unity'.[1] Maxwell's electromagnetic theory of light, which unifies luminiferous and electromagnetic media, provides an illustration of Whewell's doctrine of consilience.

Given the theoretical power of this theory, it might have been anticipated that Maxwell would broaden the scope of the electromagnetic theory of light in the *Treatise*, to encompass an electromagnetic theory of optical reflection and refraction. But he did not do so; and though he gave a detailed treatment of the Faraday magneto-optic rotation, appealing to the rotation of molecular vortices in the ether (see Chapters V.2 and VI.1), the range of his optical theory remained essentially similar in its physical content to that first advanced in his earlier papers.

Maxwell explains this apparent lacuna in his work in his report for the Royal Society in February 1879 on a paper submitted by George Francis FitzGerald 'On the electromagnetic theory of the reflection and refraction of light'. FitzGerald had noted that in proposing 'a very remarkable electromagnetic theory of light', Maxwell had left 'the questions of reflection and refraction untouched'. The problem involved the formulation of appropriate boundary conditions at the surface of separation of two media, and interpreting these conditions in terms of electromagnetic variables. Reviewing FitzGerald's argument, and referring to a related discussion of the derivation of the optical laws from electromagnetic principles by H. A. Lorentz in his Leiden dissertation of 1875, Maxwell observes that

> In my book I did not attempt to discuss reflexion at all. I found that the propagation of light in a magnetized medium was a hard enough subject.[2]

In 1864, in the course of writing his Royal Society paper 'A dynamical theory of the electromagnetic field' (see Chapter VI.1), he had attempted to derive the laws of optical reflection and refraction from his electromagnetic theory of light, proposing electromagnetic analogues for the elastic variables employed in theories of the luminiferous ether. Shortly before submitting his paper, he reported on his efforts to Stokes, as Secretary of the Royal Society and as an authority on optics and theories of the luminiferous ether. He declared that 'I am not yet able to satisfy myself about the conditions to be fulfilled at the surface . . . of separation of two media'. The inherent complexity of the problem led him to exclude discussion of optical reflection and refraction from his paper: 'I think you once told me that the subject was a stiff one to the best skilled in undulations' (*LP*, **2**: 186–7).

Maxwell himself did not have command over the intricacies of the wave theory of light. Reporting to the Royal Society in July 1862 on a paper by Samuel Haughton on the reflection of polarised light, he was quick to recognise Stokes' greater expertise and knowledge of the literature on the wave theory of light (*LP*, **2**: 46–7, 50–3). Yet his creation of the electromagnetic theory of light in the early 1860s led him to investigate two central areas of ether theory.

In April 1864 he set up an 'Experiment to determine whether the Motion of the Earth influences the Refraction of Light' (*LP*, **2**: 148–53). This investigation was prompted by the experiments of Hippolyte Fizeau, who had attempted to detect the ether wind, the effect of the motion of the earth through the ether.[3] Fizeau had found that the velocity of light in a tube carrying a stream of water 'takes place with greater velocity in the direction in which the water moves than in the opposite direction', as Maxwell summarised his finding. Fizeau had explained his result in terms of Fresnel's theory of the partial drag of the ether by the earth moving through it. Maxwell aimed to investigate the matter using a different experimental arrangement, suggested by a subsequent work of Fizeau's,[4] seeking to detect the effect of the Fresnel drag of the ether on the refraction of light by a glass prism. His apparatus, constructed by the leading London instrument maker Carl Becker, consisted of an observing telescope, three prisms, and a second telescope with a plane mirror at its focus, so that after refraction through the prisms light rays would be returned along their path. He calculated the displacement of the rays which would arise if there was an effect due to the ether drag, but found that 'no displacement could be observed'. He concluded that 'the result of the experiment is decidedly negative to the hypothesis about the motion of the ether in the form stated here'.

However, the problem was not quite so straightforward. In calculating the displacement arising from the Fresnel drag, Maxwell had ignored the compensating change in the density of the refractive medium. According to Fresnel's theory, the ether and the transparent medium satisfy a continuity equation at their boundary; this has the consequence that the retardation due to the medium is not affected by the motion of the earth.[5] Stokes drew Maxwell's attention to the error when he submitted his paper to the Royal Society, and returned the paper to him (*LP*, **2**: 154). Indeed, in 1846, Stokes had himself considered the possible effect of the motion of the ether on the refraction of light, and had concluded that the motion of the ether would have no effect on refraction.[6]

Maxwell withdrew his paper in response to Stokes' criticism, but he did give an account of his work in a letter written in June 1867 to the astronomer William Huggins (*LP*, **2**: 306–11). Here he reported the result of his aborted 1864 paper, that his experiment had failed to find any effect of the motion of the earth on the refraction of light; and now pointed out that Stokes had established this conclusion, which was also supported by an earlier experiment by Arago. His experiments, 'tried . . . at various times of the year since the year 1864', had 'never detected the slightest effect due to the earth's motion'. Huggins included Maxwell's letter in a paper of his own published in 1868;[7] and Maxwell later described the experiment in his *Encyclopaedia Britannica* article on 'Ether'. He suggested that the ether could perhaps be detected by measuring the variation in the velocity of light when light was propagated in opposite directions (*SP*, **2**: 769–70). He amplified his discussion in a letter written in March 1879 to the American astronomer David Peck Todd. This letter was published after Maxwell's death (in November 1879) in the Royal Society's *Proceedings* and in the journal *Nature*,[8] and his suggestion of a double track arrangement led A. A. Michelson to undertake his famous experiments on ether drag in the 1880s.

Responding in May 1864 to Stokes' criticism of his attempt to detect ether drag, Maxwell was not sanguine about the prospects for broadening the scope of his electromagnetic theory of light. He declared that

> I am not inclined and I do not think I am able to do the dynamical theory of reflexion and refraction on different hypotheses & unless I see some good in getting it up, I would rather gather the result from men who have gone into the subject. (*LP*, **2**: 155)

He did not, in the event, let the matter rest there. In writing his paper 'A dynamical theory of the electromagnetic field' he decided to confront the issue. The evidence of this endeavour consists of a letter to Stokes in October 1864 and a related manuscript fragment. His sketchily outlined argument in the draft is based on a paper on the reflection and refraction of light published by Jules Jamin in 1860.[9] Jamin had discussed the boundary conditions that determine the oscillation of the ether at the interface between two media, following the argument advanced by James MacCullagh and Franz Neumann, rather than that of Fresnel. MacCullagh and Neumann suggested that 'the vibrations in two contiguous media are equivalent', as MacCullagh had expressed it;[10] assuming this condition, Jamin derived the optical laws, developing equations connecting the oscillations with the angles of incidence and refraction.

In his draft, Maxwell attempted to establish an electromagnetic theory of optical reflection and refraction, seeking to interpolate results drawn from the electromagnetic theory of light in Jamin's and Fresnel's expressions for the oscillations of the ether at the interface between two media (*LP*, **2**: 182–5). He equates the displacement in the ether with the 'electric displacement', one of the cardinal concepts which he had deployed in obtaining his electromagnetic theory of light (see Chapter V.2).

The conservation of *vis viva* (energy) at the interface between the two media was stated by Fresnel as well as by MacCullagh and Neumann. This is the one feature of the dynamical ether models that Maxwell could accept. In his letter of October 1864 he informed Stokes that

> I am not yet able to satisfy myself about the conditions to be fulfilled at the surface except of course the condition of conservation of energy.

Yet, Fresnel, and MacCullagh and Neumann, had derived the energy equation on different assumptions. Fresnel obtained the energy equation on the supposition that the densities in the two media were different; while MacCullagh and Neumann had supposed the equality of the density of the ether in the two media. Maxwell questions Jamin's solution, based on the MacCullagh–Neumann boundary condition of the 'equality of the motion both horizontal & vertical in the two media', and the assumption that 'no such vibrations could exist in the media unless they were of equal density'. He criticises the selectivity of the conditions and assumptions, an endemic feature of dynamical ether theories.

> Therefore the general theory, which ought to be able to explain the case of media of unequal density (even if there were none such) must not assume equality of displacements, of contiguous particles on each side of the surface.

He did not pursue this attempt to derive the laws of reflection and refraction from the electromagnetic theory of light. In the draft he failed to apply Fresnel's theory consistently, as the result of a trivial slip. He told Stokes that 'I have written out so much of the theory as does not involve the conditions at bounding surfaces and will send it to the R.S. in a week' (*LP*, **2**: 186–8). Thus his paper 'A dynamical theory of the electromagnetic field' does not include an electromagnetic theory of the conditions for the reflection and refraction of light.

Maxwell's electromagnetic theory of light did not therefore encompass the range of optical problems addressed by theories of the luminiferous ether. His limitation of the range of his electromagnetic theory of light as a general theory of optics is paralleled by his apparent failure to consider the question of the production of electromagnetic waves. He conceived the generation of light to be a mechanical rather than an electromagnetic process, a phenomenon of molecular motion in the ether.[11] The generation of electromagnetic waves was not a straightforward deduction from Maxwell's concept of molecular motion in the ether. Nor did Heinrich Hertz deduce the possibility of generating electromagnetic waves from any simple interpretation of Maxwell's theory of the electromagnetic field. Hertz was working within the framework of Helmholtzian electrodynamics, a theory of physics which in its essentials diverges radically from the suppositions of Maxwellian theory. Helmholtz's electrodynamics assumed that all physical effects could be deduced from an interaction energy between physical bodies, while Maxwell's theory aimed to reduce effects to states of the mediating field or ether. While Hertz did not seek to demonstrate the propagation of electric waves, their fabrication led him to transcend Helmholtzian electrodynamics. The concept of electromagnetic waves, and the validation of Maxwell's concept of the electromagnetic field, then became fundamental to his physics.[12]

Maxwell's concept of the relation between light and electromagnetism, as grounded on molecular motion in the ether, is illustrated by his interpretation of the Faraday magneto-optic effect. Writing to Thomson shortly before the publication of the *Treatise*, he observes that 'Faradays twist of polarized light will not come out without . . . local motion' (*LP*, **2**: 784). The relation between optics and electromagnetism is expressed in terms of the hypothesis of molecular vortices, the molecular connection between ether and matter (see Chapters V.2 and VII.1). On reviewing, in February 1879, FitzGerald's attempt to create a 'purely electromagnetic hypothesis' to explain the reflection and refraction of light, he contrasts this work with his own explanation of the Faraday effect, describing his interpretation as a 'hybrid theory, in which bodily motion of the medium is made to cooperate with electric current'. In his view, 'Faraday's phenomenon remains unexplained by the pure electromagnetic theory'; this was because 'the terms which seem to lead up to a rotation of the plane of polarization, though they enter into the preliminary equations disappear from the final ones'.[13]

Maxwell therefore places emphasis on the dynamical basis of his theory of the electromagnetic field. In the *Treatise* he gives primacy to his attempt to 'construct a mental representation' of the ether through which and by which electromagnetic effects were supposed to be propagated by the transmission of energy. Having cited Gauss' letter to Wilhelm Weber, on the need to formulate a 'consistent representation' of the propagation of energy (see Chapter VI.1), in the final section of the *Treatise* he repeats Gauss' conviction 'that a theory of electric action in time would be found to be the very key-stone of electrodynamics'. He deploys Gauss' view in support of his own theory of the field:

> Now we are unable to conceive of propagation in time, except either as the flight of a material substance through space, or as the condition of motion or stress in a medium already existing in space.

He argues that energy cannot exist independently of material substances: 'whenever energy is transmitted from one body to another in time, there must be a medium or substance in which the energy exists'. The electromagnetic field is embodied in an ethereal medium: the transmission of energy thus implies 'the conception of a medium in which the propagation takes place'.

The assertion of the theory of the electromagnetic field, explicated in terms of a dynamical theory of the ether, is the special and distinctive theoretical stance of the *Treatise*, which Maxwell interprets as answering Gauss' call for a 'consistent representation' of the propagation of electrodynamic action:

> if we admit this medium as a hypothesis, I think it ought to occupy a prominent place in our investigations, and that we ought to endeavour to construct a mental representation of all the details of its action, and this has been my constant aim in this treatise.
>
> (*Treatise*, **2**: 437–8(§866))

He uses this argument, that the propagation of energy in time requires the formulation of a theory of the electromagnetic field and a dynamical theory of the ether, to contrast the merits of his own theory with the contradictions and confusions, as he presents them, in other contemporary theories of electrodynamics. He uses Gauss' authority to mount a critique of the work of contemporary German physicists.

He considers in particular the electrodynamic theories of Bernhard Riemann, Carl Neumann, and Wilhelm Weber. Riemann's paper on the subject was first presented in 1858, but published in 1867, and translated into English in the *Philosophical Magazine*. Riemann declared that his work 'brings into close connexion the theory of electricity and of magnetism with that of light and of radiant heat'.[14] Riemann had generalised Poisson's equation for electrostatic potential to obtain an equation for propagation; and as Maxwell points out, this 'equation is of the same form as those which express the propagation of waves and other disturbances in elastic media'. Though Riemann's analysis had led him to the conclusion that the actions of electric masses on each other are propagated with the velocity of light, Maxwell comments that he 'seems to avoid making explicit mention of any medium through which the propagation takes place' (*Treatise*, **2**: 435 (§862)). For this reason, Riemann's theory failed to satisfy Gauss' conviction of providing a 'consistent representation' of the propagation.

In March 1868, shortly after first reading Riemann's paper, Maxwell offered Tait an analysis of its import. He contrasts Riemann's force law with Weber's law of electrodynamic force.

> Weber says that the electrical force depends on the distance and its 1st & 2nd derivatives with respect to t[ime].
>
> Riemann says that this is due to the fact that the potential at a point is due to the distribution of electricity elsewhere not at that instant but at times before depending on the distance.
>
> In other words potential is propagated through space at a certain rate and he actually expresses this by a partial diff eqn appropriate to propagation. (*LP*, **2**: 353–4)

However, he maintains that Riemann's theory of propagation of potential violates Newton's third law of motion. This is because the potential which had been propagated due to the position of two bodies at a former time will act on the bodies, and will augment the velocity of one of the bodies. Thus 'the system will be a locomotive engine fit to carry you through space with continually increasing velocity. See Gullivers Travels in Laputa'. He draws the conclusion that only a theory of the electromagnetic field, a theory of the electromagnetic medium, can satisfactorily resolve the difficulty:

> Riemann's action & reaction between the gross bodies are unequal and his energy is nowhere unless he admits a medium which he does not do explicitly. (*LP*, **2**: 354–5)

Maxwell repeated this critique in his 1868 'Note on the electromagnetic theory of light', there directed at the argument of a paper by the Danish physicist L. V. Lorenz, as well as against Riemann (*SP*, **2**: 137–8). As Maxwell did not fail to note in his letter to Tait, this critique of Riemann pointed up the key feature of his own field theory: that energy is propagated in the ether.

> My action & reaction are equal only between things in contact not between the gross bodies till they have been in position for a sensible time, and any energy is and remains in the medium including the gross bodies which are among it.

A theory of the propagation of energy therefore demands the formulation of a field theory, where 'space contains a medium capable of dynamical actions' (*LP*, **2**: 354–5). In the *Treatise* he expounds his own dynamical theory of the electromagnetic field, which he presents as meeting Gauss' demand for a 'consistent representation'.

He also considers Carl Neumann's theory of electrodynamics.[15] He had outlined his critique in a letter to Thomson in 1869.

> C. Neumanns theory of the transmission of Potentials is altogether unique, the Potential mm_1/r (not the potential function m/r) starts from m (with the consciousness of the value of r and m_1 at the instant) and travels along r with uniform velocity not absolute but relative to m till it reaches m_1 which receives it after a time t. Truly those who supposed that Neumanns potential travelled like light were greatly mistaken.
>
> (*LP*, **2**: 499–500)

The value of the 'potential' depends on the emitting and receiving particles, and on the distance between them at the instant of emission. In the *Treatise* he observes that Neumann's theory of the 'transmission of potential' differed significantly from Riemann's representation of the action of electric masses on each other. Whereas Riemann had conceived such action to be analogous to the propagation of light, there is 'the greatest possible difference between the transmission of potential, according to Neumann, and the propagation of light'. Neumann envisaged 'potential' to be transmitted like a projectile, being constant relative to the velocity of the emitting particle at the instant of emission, rather than to the ether or space in which it is propagated. Maxwell rests his case against Neumann on his inability to construct a 'consistent representation' of the transmission of 'potential' (*Treatise*, **2**: 436 (§863)).

The most famous and long-established of all German theories of electrodynamics was that due to Wilhelm Weber, first proposed in 1846. Here Maxwell had long echoed Helmholtz's criticism of Weber's theory, that the velocity-dependent terms in Weber's force law were in conflict with the principle of the conservation of energy (*SP*, **1**: 208, 527).[16] But in November 1871 Maxwell confessed to Tait that 'Weber has reason', and that in his theory of electrodynamics 'Conservation is conserved'. However he was quick to note Helmholtz's recent critique of Weber:

> Helmholtz has shown . . . that it is possible (by Webers Law) to produce in a material particle carrying electricity an infinite velocity in a finite space and finite time and it appears from the formula that

> forthwith it is hurled with this ∞ velocity into a region where by the formula the velocity is $\sqrt{-1}$. (*LP*, **2**: 686)

Weber's electrodynamic force law, the expression of an action at a distance theory of electrodynamics, led to an absurd conclusion. In the *Treatise* Maxwell summarises the Helmholtz–Weber debate, favouring Helmholtz's position. He criticises Weber's theory as implying that particles 'may perform an infinite amount of work', and as leading to an 'impossible result' (*Treatise*, **2**: 429–31 (§§852–4)). As an action at a distance theory of electrodynamics, Weber's theory made no attempt to provide the 'consistent representation' that Gauss considered to be the 'very key-stone of electrodynamics'.

In concluding the *Treatise*, Maxwell confronts Weber, Riemann and Neumann with the views of a natural philosopher even more exalted than Gauss: Isaac Newton. Here, as in his lecture on 'Action at a distance', delivered at the Royal Institution in February 1873, shortly before the *Treatise* was published, he appeals to a Newtonian pedigree for his rejection of action at a distance and his espousal of field theory. In adopting this rhetorical strategy he follows Faraday, who had quoted with approval Newton's third letter to Richard Bentley: that it was inconceivable that matter could interact without contact, and that it was absurd that gravity should be supposed as inherent in matter.[17] If gravity, the paradigm of a force measured by the inverse square of the distance between bodies, is a property of matter, then its direct action at a distance would be intelligible. But if gravity is not a property of matter, as Newton asserts, then the gravitational attraction of bodies across space must be mediated by some agent. By analogy, Maxwell maintains that energy is mediated in the electromagnetic field by the 'medium in which the propagation takes place'.

Maxwell argues that Weber, Riemann and Neumann held 'some prejudice, or *à priori* objection, against the hypothesis of a medium in which the phenomena . . . take place' (*Treatise*, **2**: 437–8 (§§865–6)). In his lecture on 'Action at a distance', he adduces support for his own espousal of an electromagnetic medium by appealing to Newton's discussion of gravity in terms of the agency of an ether, in Query 21 of the *Opticks*. He quotes Newton's letter to Bentley, denying that 'gravity should be innate, inherent and essential to matter', acting 'without the mediation of anything else' (*SP*, **2**: 316). He maintains that the supposition of a medium in which energy is propagated is consistent with Newton's doctrines, pointing out that it was Roger Cotes, in his preface to the second edition of Newton's *Principia*, who was the author of the opinion that 'action at a distance is one of the primary properties of matter, and that no explanation can be more intelligible than this fact'. In the *Treatise* Maxwell describes this non-Newtonian doctrine as the 'dogma of Cotes'; and in February 1873 he refers to Cotes as 'one of the earliest heretics bred in the bosom of Newtonianism' (*LP*, **2**: 817). In his lecture on 'Action at a distance' he firmly declares that 'the doctrine of direct action at a distance cannot claim for its author the discoverer of universal gravitation' (*SP*, **2**: 316). He thus appeals to Newton in support of a field theory and the supposition of a mediating ether.

Writing to Faraday in November 1857, Maxwell discusses the possibility, which Faraday had raised, of extending the concept of lines of force to explain gravitation. In a paper 'On some points of magnetic philosophy' (1855), Faraday had suggested that gravity could be mediated by lines of force, thus seeking to resolve the puzzle bequeathed by Newton. The action of the sun in generating gravity was similar to the action of a magnet in generating magnetism. The 'power [of gravity] is always existing around the sun and through infinite space, whether secondary bodies be there to be acted on by gravitation or not', he wrote.[18] In a subsequent paper 'On the conservation of force' (1857), he perceived a problem in this proposed generalisation of the theory of lines of force. Here he discussed gravity in terms of the '*creation* of power' and the '*annihilation* of force' when bodies approach and recede from each other; and he concluded that 'the idea of gravity appears to me to ignore entirely the principle of the conservation of force'.[19]

As Maxwell explained to him, this conclusion rested on contemporary ambiguity in the meaning of the concept of 'force'. By 'conservation of force' Faraday meant the transformability and indestructibility of natural 'forces' or 'powers', a notion that lacks quantitative expression. His argument that 'force' would be created and annihilated as bodies approach and recede exposes the ambiguities of the expression 'the conservation of force'. Maxwell sets out the implication of the concepts of 'energy' and the 'conservation of energy', which were becoming established in physics in the 1850s.

> Now first I am sorry that we do not keep our words for distinct things more distinct and speak of the 'Conservation of Work or Energy' as applied to the relations between the amount of 'vis viva' and of 'tension' in the world; and of the 'Duality of Force' as referring to the equality of action and reaction. (*LP*, **1**: 549)

The terms '*vis viva*' and 'tension' were used by John Tyndall, as renditions of Helmholtz's terms *lebendige Kraft* and *Spannkraft*, in his 1853 translation of Helmholtz's memoir *Über die Erhaltung der Kraft* of 1847 (and correspond to 'kinetic' and 'potential' energy); while the expression 'the law of the conservation of energy' had been first used by Rankine in 1853.[20] Maxwell was employing terms of recent provenance.

He explains the distinction between 'force' and 'energy': 'Force is the tendency of a body to pass from one place to another', while 'Energy is the power a thing has of doing work'. Energy is a conserved quantity, so that in the explanation of gravity by the theory of lines of force, outlined so suggestively by Faraday

> we have conservation of energy . . . and besides this we have a conservation of 'lines of force' as to their *number* and total strength for *every* body always sends out a number proportional to its own mass, and the pushing effect of each is the same.

Maxwell found Faraday's ideas on gravity exciting, their geometrical imagery once again arousing his enthusiasm. If the concepts of 'energy' and 'lines of force' were disentangled, it would be possible to envisage a field theory of gravity: 'I do not think gravitation a dangerous subject to apply your methods to', he told Faraday. He explains his own thoughts on the matter in some detail.

> The lines of Force from the Sun spread out from him and when they come near a planet *curve out from it* so that every planet diverts a number depending on its mass from their course and substituting a system of its own so as to become something like a comet, *if lines of force were visible.*

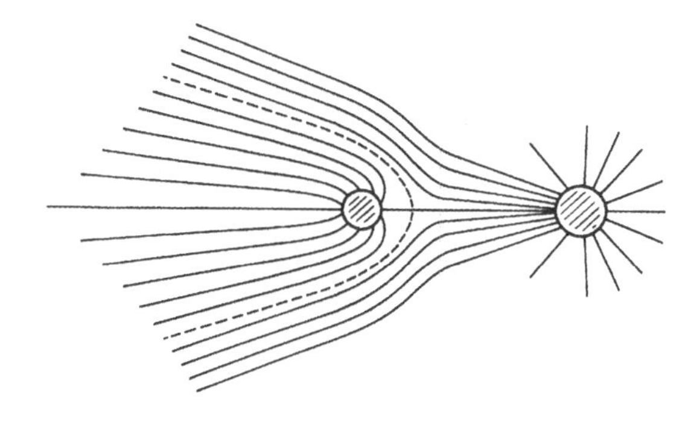

> The lines of the planet are separated from those of the Sun by the dotted lines. Now conceive every one of these lines (which never interfere but proceed from sun and planet to infinity) to have a *pushing* force, instead of a *pulling* one and then sun and planet will be pushed together with a force which comes out as it ought proportional to the product of the masses & the inverse square of the distance.
>
> (*LP*, **1**: 549–52)

The imagery of lines of force curving out from the sun, envisioned as analogous to the spreading tails of comets, remained a theme in Maxwell's continuing speculations about gravity. Writing to the Harvard astronomer George Phillips Bond in August 1863, he develops the argument of his 1857 letter to Faraday in mathematical terms. The equation of the lines of force implied 'the appearance of a comets tail being more like a catenary than a parabola near the head'. Still intrigued by the geometrical imagery of gravitational lines of force, he was now more cautious about the analogy between gravitational lines of force and the tails of comets: 'I think that visible lines of gravitating force are extremely improbable, but I never saw anything so like them as some tails of comets'.

He was now drawn to consider the question of a field theory of gravity in terms of a mechanical theory of the ether, following the transformation in the scope of his field theory in his paper 'On physical lines of force' of 1861–2 (see Chapter V.2).

> We have . . . no knowledge of the ultimate *strength* of the heavenly medium but it is well able to do all that is required of it, whether we give it nothing to do but transmit light & heat or whether we make it the machinery of magnetism and electricity also and at last assign gravitation itself to its power.

He thus envisages a generalisation of his ether theory, in which the gravitational attraction between two bodies would be explained by 'lines of pressure in the medium', incorporating gravity within the unification of optics and electromagnetism that he had claimed in 1862. But the problem with this universal ether theory, whereby 'gravitation would be explained on mechanical principles', was to explain how a body could produce and sustain a 'linear pressure radiating out in straight lines from the body' (*LP*, **2**: 106–7). The field theory of electromagnetism, which had incorporated optics, could not subsume gravity.

Maxwell discusses the question in a 'Note on the attraction of gravitation' in his paper 'A dynamical theory of the electromagnetic field' (1865). Consonant with the strategy of this paper (see Chapter VI.1), he is now concerned with the problem of gravitational energy rather than the explanation of pressure in the ether. He draws the parallel between 'lines of gravitating force' and lines of magnetic force, but points out that gravity is always an attractive force. He concludes that the 'intrinsic energy of the field of gravitation must therefore be less wherever there is a resultant gravitational force', so that if gravity was effected by an ether permeating space, then 'every part of this medium possesses, when undisturbed, an enormous intrinsic energy' which would be diminished by the presence of gross bodies within it. He sees little point in pursuing the issue.

> As I am unable to understand in what way a medium can possess such properties, I cannot go any further in this direction in searching for the cause of gravitation. (*SP*, **1**: 570–1)

Neither the geometrical imagery of lines of force, the theory of pressure in the ether, nor the concept of the transmission of energy in the field would generate a theory of gravity. But the form of the tails of comets continued to intrigue him, and perhaps held the key to the explanation of gravitation. He pursued the issue in a letter to William Huggins in October 1868. He once again takes up the question of gravitational energy, affirming the importance of ascertaining 'the form in which the energy of gravitation exists in space'. But he admits that 'we see no way of *accounting* for the proportionality of gravitation to mass by any legitimate method of demonstration'. The tails of comets, curving away from the sun, seemed anomalous to gravitational attraction. To explain the way 'the tails of comets fly off in the direction opposed to the sun', he was even prepared to allow the possibility of gravity 'being changed into repulsion by a change of state of the matter of the tail'. This would imply that 'the comet consisted of a mixture of gravitating and levitating matter'. The truth of such speculations was, he drily admits, 'another question' (*LP*, **2**: 451–2).

VIII.2 Molecules

In a memorandum on the history of the 'Kinetic Theory of Gases', written for William Thomson in the summer of 1871 (*LP*, **2**: 654–60), Maxwell summarises his contribution to the emergent science of molecular physics in his 1860 paper 'Illustrations of the dynamical theory of gases'. He notes that his paper encompassed discussion of the transport properties of gases: the 'Theory of Internal Friction of gases . . . of diffusion . . . [and the] Theory of conduction of heat in gases' (see Chapter V.1). Using his statistical model of particle collisions, he had calculated the 'mean length of path', the mean distance travelled by a gas molecule between successive collisions, from data derived from experiments on diffusion and viscosity. Finding the values he obtained to be in agreement, 'as closely as rough experiments of this kind will permit' (as he had noted in 1860), provided crucial support for his theory (*LP*, **1**: 660).

But Clausius pointed to errors in Maxwell's discussion of thermal conductivity, forcing him to review the theory. In his 1871 memorandum Maxwell concedes that Clausius' objections to his account of diffusion and the conduction of heat in gases 'were well founded'. Indeed, he compliments Clausius, noting that 'in his paper on Conduction Clausius greatly advanced the methods of treatment, and caused me to go through the subject still in the old style but improved', adding that this work was 'Not published' (*LP*, **2**: 657–8).

In 'Illustrations of the dynamical theory of gases' Maxwell explains viscosity by the transfer of momentum between successive layers of molecules, and diffusion in terms of molecular agitation and the flow of currents of gases. He supposes that heat is conducted when molecules from hot regions of gas collide with those in cooler regions, energy being transferred. In his paper 'On the conduction of heat by gases' (1862) Clausius appraised Maxwell's treatment of diffusion and thermal conductivity (*SP*, **1**: 392–405). Maxwell had assumed that the molecules of the gas would move equally in all directions; but Clausius pointed out that he had therefore disregarded the additional momentum associated with motion in the direction of the temperature gradient in a gas. In representing the distribution of molecular velocities among gas molecules, Maxwell had supposed that the distribution function is isotropic; and Clausius pointed out that this procedure implied that the conduction of heat would be accompanied or occasioned by a flow of the molecules of the gas.[21]

Maxwell drafted a response to meet this critique (*LP*, **2**: 72–85), seeking to allow for the difference in the properties of a gas along a temperature gradient. In place of representing the distance travelled by a molecule between successive collisions by the 'mean length of path' between collisions, a concept which had been introduced by Clausius (see Chapter V.1), he supposes a variable path length. This variable path length is dependent on the position of a particle between collisions, and enables variations in molecular collisions to be taken into account when the properties of a gas vary from place to place, as in the case of the conduction of heat. The calculations however proved inconclusive, and he abandoned the attempt to meet the difficulty by modifying the argument of his 1860 paper.

His decision to undertake an experimental investigation of transport phenomena may have been prompted by this failure to resolve Clausius' criticisms. The form of the apparatus used in his experimental study of gas viscosity was suggested by the method used to record the deflection of a magnet in the experiments on the standard of electrical resistance (see Chapter III.3). Writing to Stokes in June 1863, at the time of the first series of the experiments on the standard of resistance, he remarks that

> I have been studying oscillations of magnets by aid of mirrors and I hope to apply the principle to the determination of gaseous friction by means of a disc oscillating in a gas and determining the log. decrement of oscillation and the time of oscillation. (*LP*, **2**: 96)

He began his first experiments on the viscosity of gases shortly afterwards, in November 1863 (*SP*, **2**: 25);[22] but it is clear from his correspondence that systematic work, using the apparatus described in his paper on gas viscosity, which was read as the Royal Society's Bakerian Lecture in February 1866, was established in spring 1865 (*LP*, **2**: 202, 212, 214–25). The apparatus (Fig. VIII.1) used magnets to vibrate discs which were torsionally suspended in a container containing air or other gases, and the oscillations of the discs were measured by mirrors attached to the suspension (*SP*, **2**:3–7; *LP*, **2**: 231).

The experiments on gas viscosity had important implications for the molecular model assumed in the kinetic theory of gases. In his paper 'Illustrations of the dynamical theory of gases' Maxwell had established the 'physical analogy' between 'the laws of motion of an indefinite number of small, hard, and perfectly elastic spheres acting on one another only during impact' and a system of gas molecules (*SP*, **1**: 377–8). This model implied that the velocity of a gas would vary as the square root of the absolute temperature (*SP*, **1**: 389), and this was the result that he expected to confirm in his experiments in April 1865 (*LP*, **2**: 219). He found however that the viscosity was a linear function of the absolute temperature of the gas (*LP*, **2**: 224, 232), calling into question the hypothesis that gas molecules could be represented as elastic, spherical particles.

In suggesting his 'physical analogy' of elastic, spherical particles as a theory of gas molecules in 'Illustrations of the dynamical theory of gases', he had noted that the molecules could be represented in less hypothetical terms:

'instead of saying that the particles are hard, spherical, and elastic, we may if we please say that the particles are centres of force' (*SP*, **1**: 378). In his paper 'On the dynamical theory of gases' (1867), presented to the Royal Society in May 1866, he follows this approach as the more appropriate in the formulation of a theory of gases. He explains that he had abandoned the concept of gas molecules as 'hard elastic bodies acting by impact' (*LP*, **2**: 280), 'in consequence of . . . my experiments on the viscosity of air at different temperatures'.

> I propose to consider the molecules of a gas . . . as small bodies or groups of smaller molecules repelling one another with a force whose direction always passes very nearly through the centres of gravity of the molecules, and whose magnitude is represented very nearly by some function of the distance of the centres of gravity.

The experimental result that the viscosity is proportional to the absolute temperature implied that the law of repulsion between gas molecules varied 'inversely as the *fifth* power of the distance' (*SP*, **2**: 29).

This revision of his kinetic theory of gases involved a reconstruction of the concept of molecularity assumed in his theory of gases. Gas molecules are no longer considered as 'elastic spheres of definite radius', but are now defined in terms which avoid assumptions about the nature of matter: the 'molecules of a gas in this theory are those portions of it which move about as a single body'. This leaves open the question of the physical nature of the molecules. 'These molecules may be mere points, or pure centres of force endowed with inertia', he writes, or, 'if necessary, we may suppose them to be small solid bodies of a

Fig. VIII.1. Maxwell's apparatus, described in 'On the viscosity or internal friction of air and other gases' (1866), used in his experiments for determining the viscosity of gases as a function of temperature and pressure. The experiment consisted in observing the oscillations of glass discs *f*, *g*, *h* suspended in a sealed chamber between circular plates of glass *F*, *G*, *H*, *K*. The pressure of the gas in the chamber could be varied, and was read on the barometer *ACB*. The temperature of the gas could be varied by filling the tin vessel (labelled Fig. 10) with hot water, steam, or cold water, and raising it to the level of the chamber; and the temperature was read on the thermometer *T*. The suspension-piece *a* holds the suspension-wire, to which is attached the axis *cdek* held by a clip at *c*. By attaching a small piece of magnetised steel wire *ns* to the axis to which the glass discs are attached, and placing magnets under *N*, Maxwell set the discs in motion. A plane mirror *d* is attached to the axis, by which its angular position is observed through the window *C*. From *SP*, **2**: Pl. IX facing p. 24. (A photograph of the apparatus is reproduced in *LP*, **2**: Pl. IV facing p. 230.)

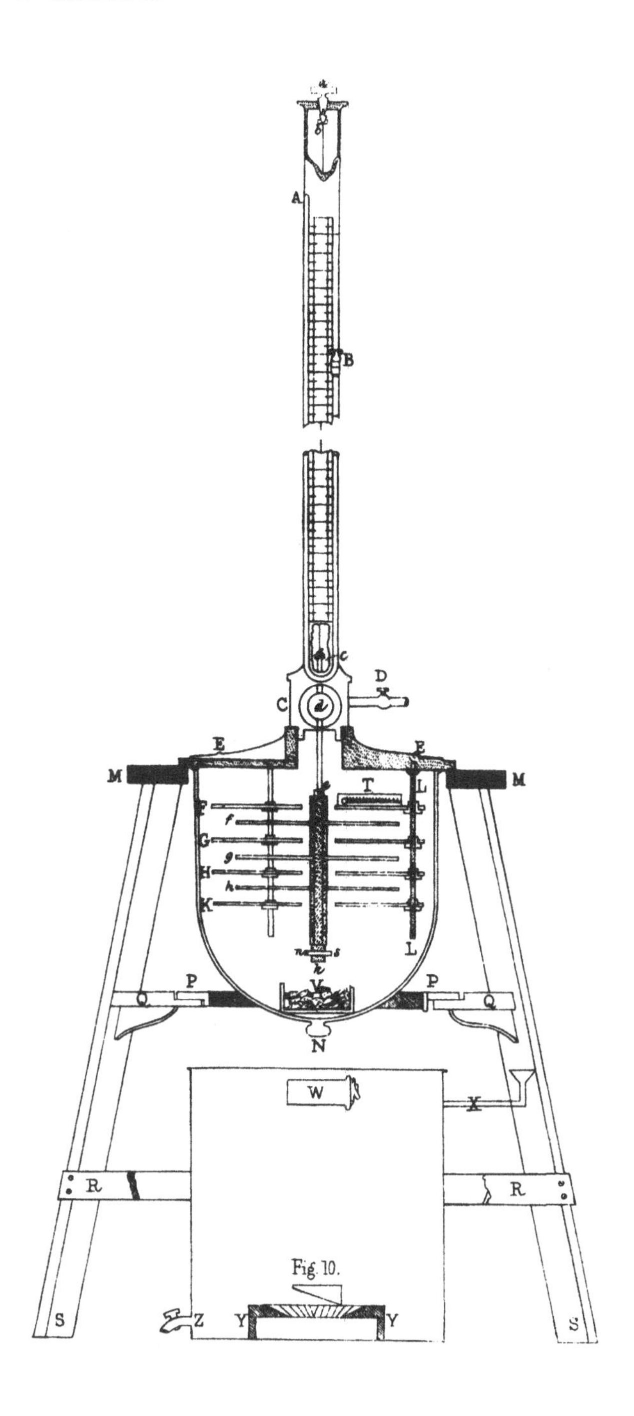
A
B
C
D
E
E
M
M
T
L
F
f
G
g
H
h
K
L
P
P
Q
Q
V
N
W
X
R
R
Fig. 10.
S
Z
Y
Y
S

determinate form' (*SP*, **2**: 29, 33). But his kinetic theory of gases is now formulated in such a way that, for the purposes of this theory, the question need not be raised. Molecules are conceived as portions of matter moving as single bodies, repelling by a force inversely proportional to the fifth power of the distance between their centres of gravity. Assumptions about molecularity are pared away: molecular interactions are conceived as encounters rather than as collisions of elastic spheres; the notion of the 'mean length of path' between collisions, and its calculation from viscosity data, is abandoned. He shows that the inverse fifth power law is consistent with explaining gas viscosity in terms of the 'time of relaxation' of stresses in the gas, a concept related to the macroscopic elastic properties of the gas (*SP*, **2**: 30–2).

In his 1873 lecture on 'Molecules', Maxwell divides the data of molecular science 'into three ranks, according to the completeness of our knowledge of them'. The first rank constituted the relative masses of molecules of different gases and the average molecular velocities, data obtained from measurements of the pressure and densities of gases. These, he declares, 'are known to a high precision'. In the second rank he places the relative sizes of molecules, their mean free paths and collision frequencies; these quantities could only be ascertained approximately until the methods of experimentation were improved. In the third rank he places the properties of molecules which might perhaps be obtained from work in molecular science: the absolute mass and diameter of molecules and their number in a given volume. Knowledge of these quantities was 'only as yet of the nature of a probable conjecture' (*SP*, **2**: 371).

From the outset of his work on the kinetic theory of gases, Maxwell had drawn attention to the implications of his theory for elucidating this third rank in his classification of molecular physics. As early as May 1859 he had told Stokes that he found that his theory had the consequence 'that equal volumes of gases at the same press. & temp. have the same number of particles' (*LP*, **1**: 610). Support for 'Avogadro's hypothesis' (as this proposition came to be known) provided an important link between the kinetic theory of gases and molecular physics and chemistry (*SP*, **1**: 390). He highlighted the importance of this result in June 1867, at a meeting of chemists sceptical of the chemical atomic theory: 'In order to decide with certainty on the truth or falsehood of the atomic theory, it would be necessary to consider it from a dynamical point of view' (*LP*, **2**: 305), that is, from the perspective of the kinetic theory of gases. In a sketch on the 'Classification of the Physical

Sciences', written in late 1872 or early in 1873, an outline of topics to be included in the ninth edition of the *Encyclopaedia Britannica,* he appended a supplementary note commenting that 'Dynamical Science is continually reclaiming large tracts of good ground from . . . Chemistry' (*LP*, **2**: 782). To ignore the implications of the kinetic theory of gases for chemistry, he warned his audience in 1867, would be to leave out of account 'considerations derived from the theory of heat' and the 'kind of dynamics treated of in books on mechanics' (*LP*, **2**: 305). For Maxwell, conclusions derived from a dynamical theory, such as the kinetic theory of gases, had especial authority (see Chapter VI.1), and were not to be readily discounted.

Shortly before delivering his lecture on 'Molecules' he had made a further advance in molecular physics, developing Joseph Loschmidt's attempt to gain an estimate of molecular diameters from a study of the diffusion of gases; Maxwell obtained values which were in agreement with calculations based on viscosity (*SP*, **2**: 343–50). He calculated the value of the molecular diameter of hydrogen as 5.8×10^{-10} metres, about five times the modern estimate; and estimated the number of molecules in a standard volume of gas (Loschmidt's number) within 30% of the modern value.[23]

Maxwell wrote a third substantial paper on the kinetic theory of gases and molecular physics, 'On stresses in rarified gases arising from inequalities in temperature', presented to the Royal Society in 1878 and enlarged in May and June 1879 (*SP*, **2**: 681–712). The paper had its origin in his interest in the experiments of William Crookes. In 1873 Crookes reported the astonishing discovery of a 'repulsion resulting from radiation'.[24] This was of especial interest to Maxwell, for in his *Treatise on Electricity and Magnetism,* published that year, he had conjectured that rays from an electric lamp 'falling on a thin metallic disk, delicately suspended in a vacuum, might perhaps produce an observable mechanical effect' (*Treatise*, **2**: 392 (§793)). Refereeing the first of Crookes' papers on the subject, in February 1874, he noted that Crookes' 'discovery, if it really indicates a repulsion due to the radiation of heat, is of... transcending scientific value', going on to remark that the effects observed by Crookes 'seem to indicate forces of much larger value' than the 'probable repulsive action of radiation' he had pointed to in the *Treatise.*[25]

In pursuing his experiments Crookes developed a device which he called a 'radiometer', composed of a partially evacuated chamber containing a paddle wheel with vanes blackened on one side and silvered on the other. When exposed to radiant heat the wheel span rapidly. Maxwell himself witnessed some of Crookes' early experiments, performed with simpler apparatus, writing to Tait in April 1874 that 'Crookes experiments . . . whip spirits all to pieces. A candle at 3 inches acts on a pith disk as promptly as a magnet does on a compass needle'.[26] He became familiar with the development of the debate on the phenomenon through refereeing Crookes' work and papers on the subject by Osborne Reynolds and Arthur Schuster, and he knew the various explanations that had been advanced and discarded to explain the radiometer effect. By 1876 it became clear that this was not a phenomenon of light pressure (discussed by Maxwell in the *Treatise*), or due to the absorption of radiation by the blackened surfaces of the vanes, nor any 'repulsion resulting from radiation', but that it arose from the action of a gas in the radiometer chamber and the temperature difference between the vanes.

In his paper 'On stresses in rarified gases', concerned with the discussion of the effect of temperature inequalities generating stresses in gases, Maxwell discusses the radiometer. To explain Crookes' discovery that 'forces act between solid bodies immersed in rarified gases . . . as long as inequalities of temperature are maintained', he suggested that the rotation of the radiometer vanes was produced by the 'sliding' of the molecules of the gas (in the evacuated chamber) over the surface of the vanes. In an appendix to the paper (dated May 1879), prompted by William Thomson's report on his paper and stimulated by reading, as a referee for the Royal Society, a paper by Osborne Reynolds, he gave a quantitative treatment of the slippage effect in the radiometer. Reynolds had suggested that a force would be produced by inequalities of temperature at the surface of a solid, 'tending to make the gas slide along the surface from colder to hotter places'; and by incorporating this idea Maxwell was able to enlarge his analysis of the surface conditions in gases. Consonant with his previous work on the kinetic theory of gases, he described his method, in this expansion of his work on molecular physics, as 'purely statistical' (*SP*, **2**: 685, 703–4).[27]

In his writings in the 1870s Maxwell devoted attention to the study of molecular forces. In his *Theory of Heat* (1871) he gave an account of Thomas Andrews' experiments, which established the significance of the critical temperature in the transition from the gaseous to the liquid state. Andrews maintained that these states of matter are continuous, being widely separated forms which are 'capable of passing into one another by a process of continual change'. He noted that liquefaction of gases under compression could 'scarcely be explained without assuming that a molecular force of great attractive power comes here into operation'.[28] Maxwell discussed the problem of molecular forces in his 1874 review (in the journal *Nature*) of the recent work of J. D. van der Waals, who had explained Andrews' results on the continuity of the gaseous and liquid states by appeal to a new equation of state, an equation based on Clausius' virial theorem (see Chapter VI.2). Van der Waals used the virial theorem, which relates the kinetic energy of molecules to the forces acting on them, to discuss intermolecular forces.

Maxwell perceived that van der Waals' study of intermolecular forces could have important bearing on the investigation of the properties of matter. Van der Waals had assumed the existence of short-range repulsive and long-range attractive forces between the molecules, but Maxwell contested his calculation of the effects of these forces on the virial, believing that van der Waals' equation 'does not agree with the theorem of Clausius on which it is founded'.[29] While van der Waals 'makes his molecules elastic spheres', Maxwell maintained that Andrews' experiments indicated that molecules should be considered as 'centres of force', a conclusion consonant with the stance of his paper 'On the dynamical theory of gases'. There he had refrained from hypotheses about the nature of gas molecules; but to explicate his ideas on molecular forces he now proceeds in overtly speculative terms, making reference to the theory of point atoms and forces posited by R. J. Boscovich in his *Theory of Natural Philosophy* (1758, $_2$1763).

> We have thus evidence that the molecules of gases attract each other at a certain small distance, but when they are brought still nearer they repel each other. This is quite in accordance with Boscovich's theory of atoms as massive centres of force, the force being a function of the distance, and changing from attractive to repulsive, and back again several times, as the distance diminishes. (*SP*, **2**: 412–13)

Boscovich's theory of atoms as centres of force was hardly obscure, even familiar at this time; so this citation would have provided a point of reference for Maxwell's speculations on molecular forces. He pursues the issue in his lecture to the Chemical Society in February 1875. Here he discusses the law of forces as molecules approach, but makes no explicit allusion to Boscovich on this occasion.

> As the particles approach each other the action first shews itself as an attraction, which reaches a maximum, then diminishes, and at length becomes a repulsion so great that no attainable force can reduce the distance of the particles to zero. (*SP*, **2**: 423)

He develops the argument, that molecules should be considered as centres of force rather than as elastic spheres acting by impact, in a May 1876 report for the Royal Society on a paper by Andrews on the gaseous state of matter:

> the action between the molecules is not like a collision, confined to a definite distance between the encountering molecules, but extends through a certain range. . . . As two molecules approach each other, the action between them is insensible at all sensible distances. At some exceedingly small distance it begins as an attractive force, reaches a maximum at still smaller distances and then becomes repulsive. In certain cases such as that of two kinds of molecules which can enter into chemical combination but which do not so combine when simply mixed, we must admit that within the region of repulsion there is a second region of attraction, and if we continue to believe that two bodies cannot be in the same place, we must also admit that the force becomes repulsive, and that in a very high degree, when the atoms are as near together as is possible.
>
> These attractive and repulsive forces may be regarded as facts established by experiment, like the fact of gravitation, without assuming either that they are ultimate facts or that they are to be explained in a particular way.[30]

Whatever their interpretation might be, the existence of these forces should be regarded as being established by Andrews' experiments.

In his paper 'On the dynamical theory of gases' Maxwell had questioned the assumption that conclusions derived from experiments on observable bodies could be considered as applicable to molecules.

> The doctrines that all matter is extended, and that no two portions of matter can coincide in the same place, being deductions from our experiments with bodies sensible to us, have no application to the theory of molecules. (*SP*, **2**: 33)

He was here expressing caution about the danger of speculating about the physical nature of gas molecules, but the argument raised issues that he pursued in his famous article on the 'Atom' (1875), published in the ninth edition of *Encyclopaedia Britannica*. Here he discusses the problem in the context of an account of Boscovich's theory of atoms.

> Boscovich himself, in order to obviate the possibility of two atoms ever being in the same place, asserts that the ultimate force is a repulsion which increases without limit as the distance diminishes without limit, so that two atoms can never coincide. But this seems an unwarrantable

> concession to the vulgar opinion that two bodies cannot co-exist in the same place. This opinion is deduced from our experience of the behaviour of bodies of sensible size, but we have no experimental evidence that two atoms may not sometimes coincide.

He also contests 'the opinion that all matter is extended in length, breadth, and depth'. If this 'prejudice', which is 'based upon our experience of bodies consisting of immense multitudes of atoms', is called into question then atoms need not have 'the so-called property of Impenetrability, for two atoms may exist in the same place' (*SP*, **2**: 448–9).

He is here arguing that there is a disjunction between the properties observed in material bodies, and those properties which can legitimately be ascribed to the atoms and molecules of which they are held to be composed. His unwillingness to allow impenetrability as an essential property of matter was however long-standing. In a lecture to his class at Marischal College, Aberdeen in November 1856, on the 'Properties of Bodies', he had noted that

> There is another geometrical property . . . which was considered of great importance by the metaphysical philosophers. I mean the so called property of the Impenetrability of Matter or the impossibility of two material things occupying the same space. I cannot at present explain the reasons which may be brought against the necessary truth of this dogma. (*LP*, **1**: 434)

His allusion to 'metaphysical philosophers' suggests his familiarity with philosophical debates over the definition of matter. In questioning the doctrine that unobservable atoms and molecules possessed the property of spatial extension, and in refusing to ascribe impenetrability to atoms, Maxwell was implicitly denying Newton's third rule of philosophising, stated in the second (1713) edition of *Principia.* Newton there declares that the qualities of 'extension, hardness, impenetrability, mobility and inertia' were universal properties of all bodies; and he argues that these properties were known through experience. He justified the ascription of these properties to the imperceptible 'indivisible particles' (atoms) of bodies by appeal to the 'analogy of nature', and declares that this is the 'foundation of all philosophy'.[31]

Maxwell regards this Newtonian metaphysical principle as philosophically 'vulgar' (*SP*, **2**: 448). Based on an appeal to sensory experience, the Newtonian doctrine of the 'analogy of nature' claims to characterise the properties of an invisible realm of atoms, primordial particles which are in principle beyond the evidence of experience. Maxwell therefore concludes that there are no good grounds for asserting that molecules and atoms can be conceived as extended, impenetrable entities. By implication, he rejects Newton's 'foundation of all philosophy', the 'analogy of nature'.

There can be little doubt that Maxwell, who was philosophically sophisticated and notably well read, was perfectly aware of the Newtonian pedigree of the doctrine he so disparages. In his *Philosophy of the Inductive Sciences* (1840, $_2$1847) William Whewell had regarded the Newtonian argument as 'a mode of reasoning far from conclusive'. He declared that the assumption that atoms possess properties of hardness and spatial extension was 'an incongruous and untenable appendage to the Newtonian view of the Atomic Theory'.[32] Maxwell was familiar with Whewell's writings (see Chapter II.3), and this discussion may have helped shape his comments in his 1856 lecture at Marischal College.

These philosophical arguments bear on Maxwell's discussion of molecular forces, highlighting his disjunction between the properties of molecules and those of gross bodies. He comments on this theme in a draft written in 1877:

> when we come to deal with very small quantities of matter its properties begin to be different from those observed in large masses
> ... the forces which we call molecular begin to show themselves acting in a different manner from those forces which are alone sensible in their action on great masses.[33]

The work of van der Waals and Andrews on molecular forces suggested that molecules should be considered as 'centres of force'. But the evidence of spectroscopy, which indicated that molecules were capable of internal vibration, was in conflict with this conclusion.

In his *Theory of Heat* and in his article on the 'Atom' Maxwell explains the bright lines in the spectra of gases as being the result of the vibrations of the molecules: 'it is the disturbance of the luminiferous medium communicated to it by the vibrating molecules which constitutes the emitted light'. He illustrates the vibrations of a molecule by the analogy of a bell when struck and set in motion. 'This motion is compounded of harmonic vibrations of many different periods ... producing notes of as many different pitches'. The aural harmonics of the bell were analogous to the spectral lines produced by vibrating molecules. The 'centres of force' are 'no doubt in their own nature indivisible', he notes, but 'they are also, singly, incapable of vibration', as required by the evidence of spectra. Nor could molecules supposed as hard particles explain 'the vibrations of a molecule as revealed by the spectroscope' (*SP*, **2**: 463, 471).

There was a similar difficulty in reconciling the evidence of spectroscopy, which required the molecules to be capable of complex internal vibrations, and the limit on these motions imposed by the equipartition theorem of the equalisation of energy (see Chapter V.1). In 1877 he remarked on Ludwig Boltzmann's attempts to resolve the conflicting demands of spectroscopy, specific heats data, and the equipartition theorem, confessing that he remained in 'thoroughly conscious ignorance' of how to proceed.[34] But from the data of spectra he was able to adduce one important conclusion, famously expressed in his lecture on 'Molecules' in 1873. The identity of the spectra of chemical elements pointed to the uniformity of molecules of the same kind throughout the universe, and implied that molecules have a common pattern:

> the exact equality of each molecule to all others of the same kind gives it, as Sir John Herschel has well said, the essential character of a manufactured article,[35] and precludes the idea of its being eternal and self-existent.

There were 'ineffaceable characters' impressed on molecules; hence matter 'must have been created'. These conclusions on the immutability and creation of molecules were the product, Maxwell insists, of a 'strictly scientific path' of reasoning, and brought the inquirer 'very near to the point at which Science must stop'.

> Science is incompetent to reason upon the creation of matter itself out of nothing. We have reached the utmost limit of our thinking faculties when we have admitted that because matter cannot be eternal and self-existent it must have been created. (*SP*, **2**: 376–7)

With these remarks he entered current debates on 'materialism' and scientific naturalism (see Chapter IX.2).

IX Physics and metaphysics

IX.1 Matter and dynamics

In a letter of July 1874 to George Chrystal, who had attended his Cambridge lectures in 1873, Maxwell remarks that

> I am at present busy with the laws of motion and other mechanical matters and I am sore afraid that if I must choose between Mr Herbert Spencer and his opponents as to the nature of the evidence I must side with Mr Spencer . . . [for] the only proofs which are more than mere illustrations of the laws of motion are as à priori as Euclids proof that lines which make alternate angles equal will not meet.[1]

He was alluding to his work writing his text *Matter and Motion* (1876), to Spencer's claim to have given *a priori* proof of the laws of motion, and to the ensuing controversy between Tait and Spencer, which had raged in the correspondence columns of the journal *Nature* from March to June 1874.

Maxwell expressed his own view of the issue in *Matter and Motion*, where he argues that our conviction of the truth of Newton's first law of motion is strengthened by recognising that its denial could not be intelligibly conceived.

> Suppose the law to be that a body, not acted on by any force, ceases at once to move. This is not only contradicted by experience, but it leads to a definition of absolute rest as the state which a body assumes as soon as it is freed from the action of external forces. It may thus be shown that the denial of Newton's law is in contradiction to the only system of consistent doctrine about space and time which the human mind has been able to form.[2]

This construal of the laws of motion is consistent with the views of Whewell and Hopkins, that the laws of motion were suggested but not established by experiment, which Maxwell had imbibed at Cambridge (see Chapter II.3).

The controversy between Spencer and Tait flared up following a review of Spencer's books by the Cambridge mathematician John Fletcher Moulton, dismissing Spencer's views on the laws of motion. In reply Spencer invoked Thomson and Tait's *Treatise on Natural Philosophy* (1867), where he found a statement, in Whewellian style, about the 'necessary truth' of physical axioms, coexisting with the assertion that the laws of motion 'must be considered as resting on convictions drawn from observation and experiment, *not* on intuitive perception'. Happy to acquiesce to the first assertion, Spencer challenged Tait to justify this latter declaration.[3]

Tait was not the man to disdain a challenge, and fired a broadside in the columns of *Nature*. From his *Sketch of Thermodynamics* (1868) he quoted his assertion 'that Natural Philosophy is an *experimental*, and not an *intuitive* science. No *à priori* reasoning can conduct us demonstratively to a single physical truth'. Spencer responded by repeating his assertion that 'the laws of motion are à priori truths . . . [and] that no experimental proofs of them are possible'.[4]

Maxwell had himself discussed these issues privately with Tait in 1867–8, and was to pursue them in public in reviews of Tait's books in *Nature* in 1878–9. Having begun reading the draft chapters of Tait's *Thermodynamics* in December 1867 (see Chapter VI.3), he responded to Tait's scorn of *a priori* reasoning. In a comment occasioned by Tait's text and his current reading of an incompetent philosophical book, he reported sourly that he generally found metaphysical works to be 'more or less ignorant discussion of mathematical and physical principles, jumbled with a little physiology of the senses' (*LP*, **2**: 335).[5] This remark should be judged as a comment on philosophical authors and books, not as indicative of a dismissal of philosophy.

He gave a more considered judgement of the relationship between physics and metaphysics a few months later. Writing in April 1868 to the Oxford scholar Mark Pattison, he declares that

> The practical relation of metaphysics to physics is most intimate. Metaphysicians differ from age to age according to the physical doctrines of the age and their personal knowledge of them.

But he adds that

> On the other hand the effect of the absence of metaphysics may be traced in most physical treatises of the present century. (*LP*, **2**: 361)

In Maxwell's view physics and metaphysics have a close mutual relationship. The physical worldview of an age shapes metaphysical argument, and philosophical analysis bears on the formulation of physical theories. In pointing to the absence of metaphysics in contemporary works on physics, he is alluding to philosophical naivety, where unexamined philosophical presuppositions masquerade behind claims to dispense with metaphysics altogether. Like Whewell (see Chapter II.3), Maxwell regards such spurious claims as the effect of bad metaphysics. He may well have had Tait's *Thermodynamics* and Thomson and Tait's *Treatise on Natural Philosophy* in mind.

When he reviewed the second (1877) edition of Tait's *Thermodynamics* in *Nature*, Maxwell commented on Tait's unequivocal assertion that there are four categories defining physical reality – force, matter, position, and motion – to one of which 'every distinct physical conception must be referred'. It was to the 'obtrusive antinomies' of Tait's 'vigorous mind' that Maxwell attributes the naivety which enabled him to make this declaration – hardly validated by experimental evidence to which he so vehemently appeals as the only source of scientific knowledge – while at the same time denouncing 'all metaphysical methods of constructing physical science'. Moreover, Maxwell concludes in astonishment, 'before we have finished the page we are assured that heat does not belong to any of these four categories, but to a fifth, called energy'. This, as Maxwell wryly notes in his amused but barbed assessment of his friend's style, was hardly the mode of argument characteristic of 'the featureless consistency of a conventional philosopher' (*SP*, **2**: 661).

Writing to Tait in July 1868 he enclosed 'a few remarks in pencil on chap II of T & T'' (*LP*, **2**: 392). In these notes he criticises the definition of matter, as '*that which can be perceived by the senses*', given in Thomson and Tait's *Treatise on Natural Philosophy*.[6] In response Maxwell declares that 'Matter is *never* perceived by the senses', and went on to cite a passage in Torricelli's *Academic Lectures*, which he recollected as being quoted by the philosopher George Berkeley in his work *De Motu* (*LP*, **2**: 395). He had accurately quoted Berkeley's text in writing to Mark Pattison the previous April:

> Berkeley quotes (with disdain) a passage of Torricelli which seems appropriate. 'Matter is nothing but an enchanted vase of Circe, which serves for a receptacle of the force and the momenta of impulse. Power and impulse are such subtle abstracts, are quintessences so refined, that they cannot be enclosed in any other vessels but the inmost materiality of natural solids.'

He annotated this text to give contemporary analogues of the terms 'power' and 'impulse' ('forza' and 'impeto' in Torricelli's original), as 'Energy' and 'Momentum? in modern language', respectively (*LP*, **2**: 365).[7] He here quotes accurately from a contemporary translation of Berkeley's *De Motu*,[8] but in jotting down his comments on Thomson and Tait's *Treatise on Natural Philosophy* he quoted 'from memory'.

> Matter is *never* perceived by the senses. According to Torricelli, quoted

> by Berkeley 'Matter is nothing but an enchanted vase of Circe fitted to receive Impulse and Energy, essences so subtle that nothing but the inmost nature of material bodies is able to contain them'. (*LP*, **2**: 395)

In this free rendition of the Torricelli text, which he subsequently took the trouble to consult and transcribe,[9] he emphasises the thrust of his objection to Thomson and Tait's lack of philosophical sophistication, 'the effect of the absence of metaphysics', as he expresses it to Pattison. According to empiricist epistemology, objects are known through their perceived effects; thus Thomson and Tait claim that matter is known through sensory perception. Maxwell contests their application of this mode of reasoning to the science of dynamics. He argues that in the science of dynamics 'matter' is not to be represented as a substance defined by properties such as extension, solidity and impenetrability, and through these properties the object of perception. 'Matter', as conceived in dynamics, is to be defined in terms of energy and momentum.

In the *Treatise on Electricity and Magnetism* (1873) Maxwell gives a definition of matter appropriate to the science of dynamics. He there defines the 'fundamental dynamical idea of matter', as being 'capable by its motion of becoming the recipient of momentum and of energy' (*Treatise*, **2**: 181 (§550)). In a manuscript written *circa* 1873 he expounds this view in relation to his construction of Torricelli's argument.

> Torricelli . . . has expressed the relation between the idea of matter on the one hand and those of force and momentum on the other neither of which can exist without the other. (*LP*, **2**: 812)

In Maxwell's dynamical theory of physics, 'matter' is considered in relation to energy and momentum, not as the object of perception.

In the closing paragraph of the *Treatise* he again quotes from the same passage in Torricelli's *Academic Lectures*, using terms which emphasise his own interpretation of its meaning:

> energy, as Torricelli remarked, 'is a quintessence of so subtle a nature that it cannot be contained in any vessel but the inmost substance of material things'.

He deploys this statement to affirm that the transmission of energy between bodies could only be conceived in terms of 'a medium or substance in which the energy exists'. This expresses the cardinal conceptual feature of his theory of the distribution of energy in the electromagnetic field: 'the conception of a medium in which the propagation takes place' (*Treatise*, **2**: 438 (§866)). Energy could only exist in 'matter'; indeed 'matter' is defined in terms of its capacity to act as a receptacle for energy; and the propagation of energy in the electromagnetic field implies that energy is distributed through the mediation of a material substance, the 'medium' of his electromagnetic theory of light (see Chapters VI.1 and VIII.1).

Unmoved by Maxwell's remonstrance, Thomson and Tait retained their definition of matter in the second edition of their *Treatise on Natural Philosophy*, published in 1879; and on this occasion Maxwell criticised them publicly on reviewing the book in *Nature* (*SP*, **2**: 776–85). In his review he describes their definition of matter, 'as *that which can be perceived by the senses, or as that which can be acted upon by, or can exert, force*',[10] as a 'pusillanimous statement'; he is unequivocally contemptuous. The condemnation is heightened by his allusion to Tait as the co-author of *The Unseen Universe* (1875) and *Paradoxical Philosophy* (1878), works urging a theological interpretation of current physics, where it is claimed that the law of the conservation of energy is consistent with the Christian doctrine of immortality (see Chapter IX.2). He describes Tait, with some sarcasm, as an author who 'never misses an opportunity of denouncing metaphysical reasoning, except when he has occasion to expound the peculiarities of the Unconditioned'.

Maxwell uses philosophical analysis to demonstrate the hidden metaphysical assumptions which lurk behind Thomson and Tait's appeal to empirical principles. For his part, he espouses a dynamics which is inherently non-metaphysical. He points to their implicit appeal to a metaphysics based on an empiricist epistemology; this lies behind their definition of 'matter' as 'that which can be perceived by the senses'. He argues that this is a metaphysical definition of matter, and is 'very much out of place in a treatise on Dynamics'. The basic concepts of dynamics are to be defined in a manner which is strictly independent of any philosophical speculations about the substratum of material reality. The science of dynamics did not consider 'matter' in this sense. 'Real bodies may or may not have such a substratum'; this issue is metaphysical, not a question to be raised in the science of dynamics, for dynamics is concerned with the concept of 'mass'. He declares that

> Every body and every portion of a body in dynamics is credited with a certain quantitative value, called its mass.

The concept of 'matter' as employed in dynamics was not to be defined as the object of sensory perception (as Thomson and Tait supposed), or as an unknown substratum of bodies; 'matter' in dynamics denotes the dynamical concept of 'mass'.

> We must be careful, however, to remember that what we sometimes, even in abstract dynamics, call matter, is not that unknown

> substratum of real bodies, against which Berkeley directed his arguments, but something as perfectly intelligible as a straight line or a sphere.

Maxwell here alludes to Berkeley's critique of the ascription of primary qualities of extension and solidity to the unobservable corpuscles which natural philosophers such as Boyle and Newton had postulated as the constituents of physical bodies. Using reasoning that Maxwell rejects (see Chapter VIII.2), Newton had justified his argument by appeal to the 'analogy of nature', claiming that the properties of the unobservable atoms, the primary constituents of substances, could be inferred from the properties of bodies perceived by the senses. Maxwell's point is that the notion of 'matter' conceived as the material substratum of bodies is not considered in the science of dynamics.

> Whatever may be our opinion about the relation of mass, as defined in dynamics, to the matter which constitutes real bodies, the practical interest of the science arises from the fact that real bodies *do* behave in a manner strikingly analogous to that in which we have proved that the mass-system of abstract dynamics *must* behave.

Nor was he satisfied with Thomson and Tait's exposition of the concept of inertia, citing their definition.

> Matter has an innate power of resisting external influences, so that every body, so far as it can, remains at rest or moves uniformly in a straight line.[11]

This definition was drawn from 'Definition III' of Newton's *Principia*:

> Force innate in matter is the power of resisting whereby each individual body, inasmuch as it is in it to do so, perseveres in its state of resting or of moving uniformly straight on.

Newton further clarifies this definition: 'the innate force of matter may be called, by a very significant name, the force of inertia [*vis inertiae*]'.[12]

In mounting his critique, Maxwell proceeds as before. He denies that 'matter', as a dynamical concept, 'has any power, either innate or acquired, of resisting external influences'. He terms this Newtonian concept of matter the 'Manichaean doctrine of the innate depravity of matter' (alluding to the doctrine of the dualism of good and evil). In place of this metaphysical concept of matter, defined by an innate resisting power, he proposes an 'idea appropriate to dynamics': this is the concept of 'mass'. The concept of 'mass', rather than 'inertia', is 'appropriate' to the science of dynamics; here Maxwell evokes Whewell's doctrine of 'fundamental ideas' (see Chapter II.3). Thus in his exposition of Newton's first law of motion in his text *Matter and Motion* he makes no reference to the concept of inertia.

There is therefore a marked contrast between the accounts given by Newton (and Thomson and Tait) and by Maxwell of the conceptual foundations of dynamics. According to his third rule of philosophising (see Chapter VIII.2), inertia, along with extension, solidity and impenetrability, is held by Newton to be one of the essential or primary properties of matter. Inertia is a defining property of matter, and explains why a body perseveres in its state of rest or uniform rectilinear motion. By contrast, for Maxwell the concept of 'mass' is the 'appropriate' dynamical concept of matter; 'mass' is 'not that unknown substratum', but a concept 'as perfectly intelligible as a straight line or a sphere' (*SP*, **2**: 778–81).

The appeal to the intelligibility of geometry echoes the values of Scottish mathematics and philosophy (see Chapter II.3). The Whewellian resonances of the argument are again clearly explicit in a similar discussion in the manuscript cited above, written *circa* 1873. There Maxwell argues that as soon as we clearly understand the relation between the laws of motion and the concept of mass, then

> we have formed an idea of mass as the quantitative aspect of matter which is as necessary a part of our thoughts as the triple extension of space or the continual flux of time.

In the Whewellian sense, 'mass' is a necessary truth in that its elimination from our conceptualisation could not be conceived. He notes that a 'metaphysician' may have come to the conclusion 'that the property commonly called inertia is the fundamental and inseparable part of matter'. But this is irrelevant to the science of dynamics. The crucial issue in dynamics is that there are dynamical reasons for asserting that the mass of a body was measurable and constant, and that 'for all dynamical purposes a body must be measured by its mass' (*LP*, **2**: 811–12).

Thomson and Tait's concept of matter was therefore, in Maxwell's view, inherently metaphysical, paradoxically so, since Tait was famously vehement in trouncing all metaphysics. But, as Maxwell suggests in Whewellian vein, these errors arose from a failure to recognise lurking metaphysical assumptions, and were a consequence of naive philosophy. Thomson and Tait had therefore become ensnared by metaphysics; by contrast, as Maxwell told Pattison in April 1868, in his own construction of dynamics 'Inertia . . . means not metaphysical passivity', but mass (*LP*, **2**: 363–4).

1 Matter and dynamics

On writing to Pattison in April 1868 Maxwell raised a further element in his 'dynamical idea of matter'. He remarks that

> I cannot admit any theory which considers matter as a system of points which are centres of force acting on other similar points, and admits nothing but these forces. For this does not account for the perseverance of matter in its state of motion and for the measure of matter.
>
> (*LP*, **2**: 365)

There can be no doubt that he is here alluding to a variant of Boscovich's theory of matter as centres of force (see Chapter VIII.2). On giving an account of Boscovich's theory in his *Encyclopaedia Britannica* article on the 'Atom' (1875) he points out that the atoms of Boscovich were supposed to have a 'definite mass'; and in his lecture 'On action at a distance' (1873) he notes that Boscovich 'did not forget . . . to endow his mathematical points with inertia' (*SP*, **2**: 449, 317). But contemporary variants of the theory were not so circumspect.

He discusses the issue in the manuscript cited above, written *circa* 1873 (and which is probably a draft of the lecture 'On action at a distance'), where he notes that

> as we have no evidence as to the size and shape of . . . atoms some have thought it more philosophical to speak of them as centres of force, without attributing to them any finite extension. This would be quite legitimate, provided each centre of force is admitted to have mass . . . .
>
> (*LP*, **2**: 812)

But these features of Boscovich's theory, he notes in the published lecture, had been forgotten by 'some of the modern representatives of his school' (*SP*, **2**: 317). Here Maxwell may have had Michael Faraday in mind.

In his discussion, in the *Treatise*, of Faraday's theory of 'all space as a field of force', he alludes to Faraday's 'Speculation touching electric conduction' (1844) and his 'Thoughts on ray-vibrations' (1846). In these papers Faraday had contested the conventional atomic theory, envisaging matter as a system of 'powers' of attraction and repulsion filling space; he declares that 'the substance consists of the powers'. According to this notion of matter, 'the particle indeed is supposed to exist only by these forces, and where they are it is'.[13] Faraday replaces the Newtonian dualism of atoms and forces by a monism of force which defines the essence of matter. He deploys these ideas on matter in expounding his theory of the primacy of lines of force around 1850 (see Chapter IV.1).

In commenting on Faraday's remarks Maxwell observes that Faraday

> even speaks of the lines of force belonging to a body as in some sense part of itself, so that in its action on distant bodies it cannot be said to act where it is not.

But he goes on to claim that this notion was 'not a dominant idea with Faraday'. He interprets Faraday in the light of his own representation of the concept of lines of force. For Maxwell, the concept of lines of force expresses the geometrical structure and intensity of the electromagnetic field (see Chapter IV.2). He gives his own reading of Faraday's intentions:

> I think he would rather have said that the field of space is full of lines of force, whose arrangement depends on that of the bodies in the field, and that the mechanical and electrical action on each body is determined by the lines which abut on it. (*Treatise*, **2**: 164 (§529))

Maxwell distances his conception of space filled with a field of force, determined by the bodies in the field, from Faraday's monism of force, which had its origins in the speculative theories of matter, shaped by the ideas of Newton and Boscovich, which were current in British natural philosophy between 1700 and 1850.[14]

Maxwell was similarly cautious about modern conjectures which sought to relate space to matter. Responding to a query of Tait's in November 1874 about Riemann's non-Euclidean geometry, he declares that the 'Riemannsche Idee is not mine'. Here he is alluding to Riemann's paper, translated by William Kingdon Clifford in *Nature* in 1873,[15] and perhaps also to Clifford's paper 'On the space-theory of matter' (1870). Clifford had speculated that the motion of matter was to be understood as a 'variation in the curvature of space', presenting his conjecture as being developed from Riemann's concept of the constant curvature of space.[16] Maxwell makes the following comment:

> The Riemannsche Idee is not mine. But the aim of the space-crumplers is to make its curvature uniform everywhere, that is over the whole of space whether that whole is more or less than ∞. The *direction* of the curvature is not related to one of the $x\,y\,z$ more than another or to $-x-y-z$ so that as far as I understand we are once more on a pathless sea, starless windless and poleless totus teres atque rotundus.[17]

The Latin phrase ('all smooth and round', like a sphere, a Stoic metaphor for a wise man) is from Horace.[18]

Maxwell focuses on the problem of the definition of coordinates in Riemann's theory of constant curvature. He did not accept the curvature of space as supposed by the 'space-crumplers', presumably Riemann and Clifford. In a manuscript written *circa* 1857 'On Absolute Space' he had sought to 'render distinct the idea of space as independent of matter' (*LP*, **1**: 515), and he affirms the concept of absolute space in *Matter and Motion.*

> Absolute space is conceived as remaining always similar to itself and immovable. . . . [T]here is nothing to distinguish one part of space from another except its relation to the place of material bodies.[19]

Maxwell affirms the dualism of matter and space, and hence rejects Clifford's notion of matter as defined by the curvature of space. He envisages the field as being *in* space, rather than conceiving the curvature of the lines of force as defining the geometric structure of space. He did not find the geometric basis for a theory of lines of force in the speculations of Riemann and Clifford. He affirms the dualism of matter and field, and hence rejects Faraday's idea of matter conceived as a monism of force.

IX.2 Materialism and determinism

In his *Encyclopaedia Britannica* article on the 'Atom' (1875) Maxwell distinguishes between establishing molecular physics, based on the kinetic theory of gases (see Chapter VIII.2), and formulating a 'complete theory of matter'. By this he means a fundamental theory, and explains that

> one of the first, if not the very first desideratum in a complete theory of matter is to explain – first, mass, and second, gravitation.

He goes on to consider two possible 'complete' theories of matter, both of which he found problematic. The first is William Thomson's theory of 'vortex atoms', based on Thomson's work on vortex motion developed from Helmholtz's demonstration that vortex filaments in a perfect fluid would be immune to destruction or dissipation. Maxwell had been interested in theories of vortex motion since 1857, specifically as applied to the explanation of magnetism (see Chapter V.2), and the papers of Helmholtz and Thomson on vortex motion in a fluid had aroused his interest (see Chapter VII.2).

In the 'Atom' article Maxwell considers Thomson's development of the theory of vortex motion. He explains that Thomson's fluid is not molecular, but its motion 'renders it molecular'. Writing to Tait in March 1875, he points out that

> Thomsons Ur-fluid [original or primitive fluid] is not molecular for molecules have to be created out of it each by its own peculiar spin.[20]

The motion of the fluid converts its elements into vortex rings; and the molecules of bodies were supposed to be structured from vortex rings. The laws of motion of the 'primitive fluid' thus establish the immutable properties of atoms, conceived as vortex rings. Stating that 'it is of the essence of matter to be the receptacle of momentum and energy' (see Chapter IX.1), he concludes that Thomson's theory of vortex atoms could in principle explain mass, because 'a vortex ring at any given instant has a definite momentum and a definite energy'. But the application of the theory to explain the mass of real material bodies was problematic. According to the theory of vortex atoms, it is the mode of motion of the primitive fluid which 'renders it molecular'; and he points out that it would be 'a very difficult task' to show that bodies built from vortex rings would have a specific momentum and energy. For this reason, 'the mass of bodies requires explanation' in the theory of vortex atoms (*SP*, **2**: 472).

The second theory he considers is the theory of matter of G. L. Le Sage, published around 1800, which supposed that the gravitation of bodies is caused by the impact of 'ultramundane corpuscles', particles which emanate from beyond the physical universe (*SP*, **2**: 463). Maxwell had become aware of Le Sage's speculations on reviewing antecedents of the kinetic theory of gases (*LP*, **2**: 655), and had formed the conclusion that his 'theory of impact is faulty' (*SP*, **2**: 28).[21] He explained the basis of his criticism in the 'Atom' article: if the corpuscles were perfectly elastic spheres, their collision could not explain gravity; and if they were to rebound with a diminished velocity, they would lose kinetic energy. To salvage this difficulty, Thomson had published a paper[22] suggesting that the corpuscles could be vortex atoms 'set off into a state of vibration at impact'. But Maxwell argues that 'the impact of the corpuscles would raise all bodies to an enormous temperature', so that the system would be thermodynamically unstable. Moreover, Le Sage's theory supposed that

> the habitable universe, which we are accustomed to regard as the scene of a magnificent illustration of the conservation of energy as the fundamental principle of all nature, is in reality maintained in working order only by an enormous expenditure of external power . . . .

He concludes that Le Sage's explanation of gravity 'falls to the ground' (*SP*, **2**: 476–7).

Neither of the proposed 'complete' or fundamental theories of matter proved capable of satisfying Maxwell's desideratum: to explain mass and gravity. Such explanations were in principle possible; but in concluding his article he turned to an even more intractable problem: the 'establishment of the existing order of nature', the formation of atoms and molecules. He repeats the declaration in his 1873 lecture on 'Molecules' (see Chapter VIII.2), that the 'creation of matter itself out of nothing' lay beyond the realm of scientific explanation.

> We have reached the utmost limit of our thinking faculties when we have admitted that, because matter cannot be eternal and self-existent, it must have been created.

Science is concerned with 'the form in which [matter] . . . actually exists', not with its creation, the 'common cause or common origin' of atoms.

This doctrine of a boundary to a scientific explanation, that only the 'form' but not the 'formation' of molecules was transparent to scientific investigation, had theological implications (*SP*, **2**: 482). Maxwell had argued that there are limits to scientific explanation in his inaugural lectures at Marischal College, Aberdeen in November 1856 and at King's College London in October 1860, affirming that it was 'the peculiar function of physical science to lead us to the confines of the incomprehensible' (*LP*, **1**: 427, 670). On writing 'Molecules' and 'Atom' he developed this theme in relation to the creation of matter, giving the argument a focus on the problems of materialism and scientific naturalism that, in the wake of Darwin's theory of evolution by natural selection (published in 1859), exercised his circle in the 1860s and 1870s.

In Britain in the 1870s 'materialism' was associated with the views of the physicist John Tyndall, who in an address to the British Association for the Advancement of Science in 1868 had propounded a physical worldview which he termed 'scientific materialism'.[23] Tyndall based his concept of nature on contemporary molecular physics. Taking his place on the platform two years later, Maxwell summarised 'the penetrating insight and forcible expression of Dr Tyndall' in the following terms:

> molecules obey the laws of their existence, clash together in fierce collision, or grapple in yet more fierce embrace, building up in secret the forms of visible things.

In place of a worldview defined by doctrines of matter, Maxwell suggests a more elevated vision of scientific knowledge which would transcend Tyndall's reductionism, and which – his language echoing Whewell's (see Chapter II.3) – sought the 'still more hidden and dimmer region where Thought weds Fact' (*SP*, **2**: 216).

In his Presidential 'Belfast Address' to the British Association in 1874, Tyndall elaborated his position, evoking a dynamic concept of matter.

> Believing as I do, in the continuity of nature, I cannot stop abruptly where our microscopes cease to be of use. . . . By a necessity engendered and justified by science I cross the boundary of the experimental evidence, and discern in that Matter which we, in our ignorance of its latent powers, and notwithstanding our professed reverence for its Creator, have hitherto covered with opprobrium, the promise and potency of all terrestrial life.

While he indicated that his dynamic concept of matter evoked pantheism, and noted that the '"materialism" here professed may be vastly different from what you suppose', he was seen as the apostle of materialism.[24]

Together with Thomas Henry Huxley, Tyndall espoused a programme of explanation which has been termed scientific naturalism.[25] These men sought to detach the methods and ideology of science from theological implications. In their writings, the theologically neutral, progressive and secular optimism of the scientific endeavour was seen as expressing the 'material spirit of the age', as Maxwell succinctly put it in his inaugural lecture at Cambridge in October 1871 (*SP*, **2**: 251). Scientific naturalism was grounded on the major developments in science in the nineteenth century: the growing commitment to a belief in the uniformity of nature, in geology and in Darwin's theory of evolution by natural selection; the rejection of suppositions of divine interventions to explain discontinuities in the natural world; the restriction of divine action to the creation of the universe; and the formulation of the principle of the conservation of energy which, in Tyndall's words, brought 'vital as well as physical phenomena under the dominion of that law of causal connection which . . . asserts itself everywhere in nature'.[26] Appealing to molecular physics, Tyndall and Huxley sought to sever the ties between science and natural theology. Speaking as an authority, in his lecture on 'Molecules' and his article on the 'Atom' Maxwell contested the reductionist claims of scientific naturalism.

Tyndall and Huxley deployed scientific naturalism in advocating the professionalisation of science, seeking to project an image of science as generating positive knowledge. They detached science from the traditional language of natural theology, which had sought to emphasise the conciliation of scientific knowledge and religious beliefs and values. The emphasis on scientific progress, and the claim that science was achieving emancipation from religious orthodoxy and intellectual obscurantism, suggests links between scientific naturalism and the 'positivism' of Auguste Comte. In his 'Belfast Address' Tyndall described the 'impregnable position of science', which he characterised as striving to 'wrest from theology, the entire domain of cosmological theory'. No scheme of knowledge could 'infringe upon the domain of science', and any system of thought that did so must 'submit to its control'.[27] There was a claim here for intellectual leadership, its transference to a new cultural elite: the clerisy of the secular, scientific professional.

This positivist epistemology of science, expressed in terms of 'scientific materialism', was antipathetic to Maxwell's more traditional construction of the values and cultural implications of science. Present in Tyndall's Belfast audience, he responded with a poem, 'Notes of the President's Address', published anonymously in *Blackwood's Magazine* in November 1874, which captures Tyndall's materialism and claim to modernity:

> There is nothing but atoms and void, all else is mere whims out of date!
> Then why should a man curry favour with beings who cannot exist,
> To compass some petty promotion in nebulous kingdoms of mist?
> . . .[28]

It was in intended refutation of Tyndall's 'Belfast Address' that Balfour Stewart and Peter Guthrie Tait published (anonymously) in 1875 their book *The Unseen Universe: or Physical Speculations on a Future State.* Their vision of nature is based on the concept of energy and its transference. They maintain that there is an invisible realm which is in communication with the visible universe. Energy travelling through space is 'gradually transferred into an invisible order of things', carried though the ether, which is conceived as a 'bridge between one order of things and another'. The visible and unseen realms together constitute the 'Great Whole': 'when we assert the conservation of energy it is as a principle applicable to the whole universe', both material and immaterial. While the visible, material universe is transient, the 'Great Whole is infinite in energy, and will last from eternity to eternity'. The conservation of energy and 'the principle of Continuity (rightly viewed)' would secure immortality beyond the physical, material universe.[29] Utilising the physical and philosophical concepts deployed by Tyndall (energy conservation and the 'principle of continuity'), they use his own weapons to contest 'scientific materialism'.

In seeking to contest the claims of 'scientific materialism' Maxwell did not wish to advocate a form of scientific obscurantism. Writing to a bishop in November 1876, in response to a query about his 'Molecules' lecture, he observes that

> The rate of change of scientific hypothesis is naturally much more rapid than that of Biblical interpretations, so that if an interpretation is founded on such an hypothesis, it may help to keep the hypothesis above ground long after it ought to be buried and forgotten.

Any harmonisation between science and religion was personal, and, he wrote in March 1875, 'ought not to be regarded as having any significance except for the man himself, and to him only for a time'.[30] Unwilling to be drawn into public debate and to profess the accommodation of science and religion, he nevertheless made clear his differences with Tyndall.

He centres his argument on the rebuttal of the deterministic claims of 'scientific materialism', appealing, like Tyndall, to molecular physics, but concluding that neither the kinetic theory of gases nor the laws of thermodynamics implied materialism. Commenting on Tyndall's molecular physics in his lecture on 'Molecules', he points out that Lucretius, the exponent of ancient atomism, had allowed for free will in his materialistic system of philosophy.

> And it is no wonder that he should have attempted to burst the bonds of Fate by making his atoms deviate from their courses at quite uncertain times and places, thus attributing to them a kind of irrational free will, which on his materialistic theory is the only explanation of that power of voluntary action of which we ourselves are conscious.
>
> (*SP*, **2**: 373)

The determinism implicit in Tyndall's 'scientific materialism' did not receive support from ancient authority. But Maxwell had long expressed disagreement with determinism. On reading the first volume of H. T. Buckle's *History of Civilization in England* on its publication in 1857, Maxwell commented to Lewis Campbell on Buckle's view of history as shaped by social laws: 'a bumptious book, strong positivism, emancipation from exploded notions, and that style of thing, but a great deal of actually original matter' (*LP*, **1**: 576).

The energy concept and its implications beyond physics played a crucial role in the debates on materialism. Maxwell almost certainly attended Helmholtz's lecture at the Royal Institution in London in April 1861, on the application of the principle of the conservation of energy to organisms.[31] Helmholtz there used the energy principle to contest traditional vitalism, which held that organisms possess special vital forces which stand outside the framework of physical laws. Allowing that there may be 'other agents acting in the living body than those agents which act in the inorganic world', he maintained that these forces 'must be of the same character as inorganic forces' in that 'their effects must be ruled by necessity'. There could be no 'arbitrary choice in the direction of their actions' which were subject to physical law, the law of the conservation of energy.[32]

Responding in April 1862 to Lewis Campbell's request for information about Helmholtz's energy theory, Maxwell concurred with Helmholtz's refutation of vitalism. He drew the implication that 'the soul is not the direct moving force of the body', for if it were 'it would not last till it had done a certain amount of work, like the spring of a watch which works till it is run down'. In making explicit mention of the soul he made Helmholtz's arguments bear directly on the theological resonances of vitalism; he was, after all, writing to an ordained minister. According to Helmholtz's view of energy conservation, he explains, 'Food is the mover', supplying energy which is converted into the animating power of living creatures. In so doing, food is consumed; it 'perishes in the using, which the soul does not'.

In shaping the argument to delimit the relation between body and soul, transcending the terms of Helmholtz's paper, he seeks an appropriate analogy to represent their relation.

> There is action and reaction between body and soul, but it is not of a kind in which energy passes from the one to the other, – as when a man pulls a trigger it is the gunpowder that projects the bullet, or when a pointsman shunts a train it is the rails that bear the thrust.

Energy is not transferred from the human agent to the bullet or to the train. By analogy, the soul could act on the body without energy passing from soul to body. But the analogy was merely suggestive, and Maxwell disclaims any scientific explanation of the relation between body and soul: 'It is well that it will go, and that we remain in possession, though we do not understand it' (*LP*, **1**: 711–12).

In his review of Stewart and Tait's *Paradoxical Philosophy: A Sequel to the Unseen Universe* (1878) he again illustrates the agency of the soul by using the analogy of a guiding agent directing energy. He notes that 'the application of energy may be directed without interfering with its amount'. This could perhaps explain the action of the soul on the material bodies of living beings:

> Is the soul like the engine-driver, who does not draw the train himself, but, by means of certain valves, directs the course of the steam so as to drive the engine forward or backward, or to stop it? (*SP*, **2**: 760)

He turns the energy concept against scientific materialism.

In his *Encyclopaedia Britannica* article on 'Diffusion' (1878) Maxwell points out that the idea of the dissipation of energy depends on the limited extent of our control of nature, our ability only 'to lay hold of some forms of energy', while other forms, such as 'the confused agitation of molecules which we call heat', elude our grasp. But the notion of the dissipation of energy 'could not occur to a being . . . who could trace the motion of every molecule and seize it at the right moment' (*SP*, **2**: 646).

This recalls the 'finite being' introduced in his thought-experiment on the second law of thermodynamics: a 'finite being' is supposed capable of tracing the path and directing the motion of molecules through an aperture, so as to violate the second law of thermodynamics, without itself consuming energy in the process. He used the analogy of a 'pointsman' directing the course of a material system in giving an account of his 'finite being' in his letter of December 1870 to John William Strutt (see Chapter VI.3). He describes this 'doorkeeper' as 'a mere guiding agent (like a pointsman on a railway with perfectly acting switches who should send the express along one line and the goods along another)'. As if to forestall Strutt supposing that the 'finite being' (deliberately described thus) might be a nonnatural agent, he adds that the switches could be 'self-acting'. The thought-experiment illustrates the spontaneous fluctuations of molecules and the statistical nature of the second law of thermodynamics. He uses the analogy of a 'pointsman' to illustrate the statistical nature of the second law of thermodynamics and the agency of the soul on the body.

In his letter to Strutt he goes on to show that the perfect reversibility allowed for by laws of mechanics, that all things could 'happen backwards', is constrained by the irreversibility of natural processes described by the second law of thermodynamics: 'if you throw a tumblerful of water into the sea you cannot get the same tumblerful of water out again' (*LP*, **2**: 582–3). This argument bore on Maxwell's critique of materialism. Because proponents of materialism placed emphasis on the explanatory power of the laws of matter and motion, on molecular physics and the laws of mechanics, the irreversibility of natural processes could be shown to demonstrate the inherent limitation of the reversible laws of mechanics, and thereby to contradict the claims of philosophical materialism.

2 Materialism and determinism

Maxwell had first outlined the contrast between reversible and irreversible processes, presenting the argument in explicit refutation of materialism, in writing to Mark Pattison in April 1868 (see Chapter VI.3). He remarks that:

> A strict materialist believes that everything depends on the motion of matter. He knows the form of the laws of motion though he does not know all their consequences when applied to systems of unknown complexity.

A purely mechanical outlook would permit perfect reversibility; but natural processes are irreversible, a feature of nature described by the second law of thermodynamics.

> Now one thing in which the materialist (fortified with dynamical knowledge) believes is that if every motion great & small were accurately reversed, and the world left to itself again, everything would happen backwards . . . . The reason why we do not expect anything of this kind to take place at any time is our experience of irreversible processes . . . and this leads to the doctrine of a beginning & an end instead of cyclical progression for ever. (*LP*, **2**: 360–1)

Materialism implied eternalism, and this was consistent with the laws of mechanics. But the second law of thermodynamics implied the arrow of time, and a beginning and an end. As he noted in addressing the British Association in 1870,

> This idea of a beginning is one which the physical researches of recent times have brought home to us, more than any observer of the course of scientific thought in former times would have had reason to expect. (*SP*, **2**: 226)

The second law of thermodynamics implied the creation and end of the universe. In his lecture on 'Molecules' in 1873 he claims that the evidence of spectroscopy implied the creation of molecules: the exact equality of each molecule to all others of the same kind, as disclosed by spectra, gave each molecule 'the essential character of a manufactured article, and precludes the idea of its being eternal and self-existent' (*SP*, **2**: 376). Modern molecular physics and thermodynamics, far from validating the credentials of 'scientific materialism' as Tyndall claimed, refuted its scientific pretensions.

In his lecture on 'Molecules' Maxwell had pointed out that Lucretius had attributed free will to the undetermined swerve of his atoms. This formed an important element in his claim that the kinetic theory of gases did not imply materialism. But this was an issue that he had considered well before Tyndall's utterances on 'scientific materialism' in 1868. In the course of writing the historical preliminary to his paper 'On the dynamical theory of gases' (1867) Maxwell studied *De Rerum Natura*, and in February 1866 wrote to the Lucretius scholar H. A. J. Munro seeking advice. In his scholarly analysis of Lucretius' text, he noted that 'Lucretian atoms . . . deviate' (*LP*, **2**: 251), a point later made by Fleeming Jenkin, a member of Maxwell's circle, in his essay on 'Lucretius and the atomic theory', published in the *North British Review* in 1868.[33] In contesting the claim that ancient atomism could be used to sustain modern materialism, Jenkin argued that the action of the will would not require any transfer of energy from the system of atoms, an argument that Maxwell had used in his letter to Campbell in April 1862.

On writing to William Thomson in summer 1871 on the history of the kinetic theory of gases, Maxwell observes that Lucretius had emphasised the 'irregularity of the deflexions of the atoms introduced to account for free will &c'. He quotes directly from the text of *De Rerum Natura*, Book II, lines 292–3, on the 'minute swerving of first-beginnings' so 'the mind itself does not feel an intense internal necessity in all its actions' (*LP*, **2**: 654–5).[34]

He gave a full discussion of these issues in a paper on 'Determinism . . . and the Freedom of the Will', which he wrote in February 1873 for the Eranus club, a group of past Cambridge Apostles. Here he challenges the deterministic conclusions which materialists had drawn from the occurrence of regular, mechanical laws in nature.

> It is a metaphysical doctrine that from the same antecedents follow the same consequents. No one can gainsay this. But it is not of much use in a world like this, in which the same antecedents never again concur, and nothing ever happens twice.

The universe is fundamentally causal, yet not deterministic. He illustrates this by alluding to the swerve of Lucretian atoms, 'which at quite uncertain times and places deviate in an uncertain manner from their course'. He expounds this in terms of the instability of a dynamical system at 'singular points', invoking the example of a 'watershed, where an imperceptible deviation is sufficient to determine into which of two valleys we shall descend'. The deviation of a mechanical system at a point of singularity suggests limits to philosophical determinism and materialism:

> Our free will at the best is like that of Lucretius's atoms – which at quite uncertain times and places deviate in an uncertain manner from their course. (*LP*, **2**: 820–2)

The imperceptible deviation of a mechanical system at a 'singular point' is analogous to the action of a pointsman at a railway junction. Maxwell again uses the analogy of a pointsman in his review of Stewart and Tait's *Paradoxical Philosophy*:

> The dynamical theory of a conservative material system shews us, however, that in *general* the present configuration and motion determine the whole course of the system, exceptions to this rule occurring only at the instants when the system passes through certain isolated and singular phases, at which a strictly infinitesimal force may determine the course of the system to any one of a finite number of equally possible paths, as the pointsman at a railway junction directs the train to one set of rails or another. (*SP*, **2**: 760)

Here he refers to a recently published work on determinism and free will by the French mathematician Joseph Boussinesq. In a letter to Francis Galton in February 1879 he uses the term 'bifurcation of path' (following Boussinesq's 'lieux de bifurcation') to describe the deviation of a system at a point of singularity, so that a system will 'go off along that one of the particular paths which happens to coincide with the actual condition of the system at that instant'.[35]

As he explains in his essay on free will, these bifurcations of path at points of singularity are the result of 'potential energy, which is capable of being transformed into motion, but which cannot begin to be so transformed till the system has reached a certain configuration'. The imperceptible deviation of a system at a singular point, leading to a bifurcation of path, implied the inherent incalculability of a mechanical system subject to such instabilities.

> Much light may be thrown on some of these questions by the consideration of stability and instability. When the state of things is such that an infinitely small variation of the present state will alter only by an infinitely small quantity the state at some future time, the condition of the system, whether at rest or in motion, is said to be stable; but when an infinitely small variation in the present state may bring about a finite difference in the state of the system in a finite time, the condition is said to be unstable. (*LP*, **2**: 819, 822)

Such instabilities were not uncaused, but were incalculable. In certain cases a small error in the data would only introduce a small error in the result, but there were more complicated cases where the instabilities might prove more significant as the number of variables increases. There are therefore limits to the perfect predictability of the Laplacean deterministic universe. Tyndall's 'scientific materialism' stood in contradiction to molecular physics, thermodynamics, and the laws of motion.

Notes

The following abbreviated titles and references are used throughout the notes.

Electricity
Michael Faraday, *Experimental Researches in Electricity*, 3 vols. (London, 1839–55)

Electrostatics and Magnetism
William Thomson, *Reprint of Papers on Electrostatics and Magnetism* (London, 1872)

Life of Maxwell
Lewis Campbell and William Garnett, *The Life of James Clerk Maxwell. With a Selection from his Correspondence and Occasional Writings and a Sketch of his Contributions to Science* (London, 1882)

LP
The Scientific Letters and Papers of James Clerk Maxwell, ed. P. M. Harman, 2 vols. to date (Cambridge, 1990, 1995)

Math. & Phys. Papers
William Thomson, *Mathematical and Physical Papers*, 6 vols. (Cambridge, 1882–1911)

Natural Philosophy
William Thomson and Peter Guthrie Tait, *Treatise on Natural Philosophy, Vol. 1* (Oxford, 1867)

Phil. Mag.
Philosophical Magazine (London)

SP
The Scientific Papers of James Clerk Maxwell, ed. W. D. Niven, 2 vols. (Cambridge, 1890)

Stokes–Kelvin Correspondence
The Correspondence between Sir George Gabriel Stokes and Sir William Thomson, Baron Kelvin of Largs, ed. David B. Wilson, 2 vols. (Cambridge, 1990)

ULC
Manuscripts in the University Library, Cambridge

Chapter I Introduction: Maxwell and the history of physics

1 [P. G. Tait,] 'Clerk-Maxwell's Electricity and Magnetism', *Nature*, **7** (24 April 1873): 478–80.
2 *The Collected Papers of Albert Einstein, Volume 2*, ed. John Stachel (Princeton, 1989): 150.

3 Albert Einstein, 'Maxwell's influence on the development of the conception of physical reality', in *James Clerk Maxwell. A Commemoration Volume, 1831–1931* (Cambridge, 1931): 66–73, on 73.
4 *Math & Phys. Papers*, **2**: 34.
5 P. M. Harman, *Energy, Force, and Matter. The Conceptual Development of Nineteenth-century Physics* (Cambridge, 1982).
6 For an outline of Maxwell's physics see C. W. F. Everitt, *James Clerk Maxwell. Physicist and Natural Philosopher* (New York, 1975), based on his article on 'Maxwell' in *Dictionary of Scientific Biography*, ed. C. C. Gillispie, 16 vols. (New York, 1970–80), **9**: 198–230.
7 William Thomson, 'Notes of Lectures on Molecular Dynamics and the Wave Theory of Light' [1884], in *Kelvin's Baltimore Lectures and Modern Theoretical Physics*, ed. R. Kargon and P. Achinstein (Cambridge, Mass., 1987): 12, 111, 206.
8 Pierre Duhem, *The Aim and Structure of Physical Theory*, trans. P. P. Wiener (Princeton, 1954): 84.
9 The literature is reviewed by Daniel M. Siegel, *Innovation in Maxwell's Electromagnetic Theory. Molecular Vortices, Displacement Current, and Light* (Cambridge, 1991)
10 William Whewell, *The Philosophy of the Inductive Sciences, Founded upon their History*, 2 vols. (London, 1840), **1**: lxxi, cxiii.
11 P. M. Harman, *Metaphysics and Natural Philosophy. The Problem of Substance in Classical Physics* (Brighton, 1982).
12 *Life of Maxwell*: 431.
13 See *LP*, **2**: 6, 949. The remarks on scientific education and research, included in the portrait of 'Faraday' in *Nature*, **8** (18 September 1873): 397–9, attributed to Maxwell in *SP*, **2**: 355–60 (on pp. 355–7), were very likely introduced by the editor of *Nature* (J. N. Lockyer) as a preamble to Maxwell's portrait of Faraday. Leaving aside the question of content, this preamble includes quotations from A. W. Williamson's Presidential Address to the British Association, currently meeting in Bradford. Williamson had delivered his Address on 17 September 1873 (the day before publication of this issue of *Nature*), so that the author of the preamble would have needed advance notice of its content. The Address was published in this same issue of *Nature* (on p. 406–15), so Lockyer, as editor, would have been able to make reference to it.

Chapter II Formative influences

1 Quoted by V. L. Hilts, 'A guide to Francis Galton's *English Men of Science*', *Transactions of the American Philosophical Society*, **65** Part 5 (1975): 59.
2 *Life of Maxwell*: 60.
3 *Life of Maxwell*: 487–8 n.
4 Robert Fairley, *Jemima* (Edinburgh, 1988): 4–5
5 D. R. Hay, *Principles of Beauty in Colouring Systematized* (Edinburgh, 1845): 2, 11.
6 D. R. Hay, 'Description of a machine for drawing the perfect egg-oval . . .', *Transactions of the Royal Scottish Society of Arts*, **3** (1851): 123–7; Hay, *First Principles of Symmetrical Beauty* (Edinburgh/London, 1846): 50ff.

7 J. D. Forbes in *SP*, **1**: 2; and see *LP*, **1**: 36 note (7).

8 *Life of Maxwell*: 75–6.

9 *Life of Maxwell*: 86–7.

10 Hilts, 'A guide to Francis Galton's *English Men of Science*': 59.

11 C. G. Knott, *Life and Scientific Work of Peter Guthrie Tait* (Cambridge, 1911): 6; David B. Wilson, 'The educational matrix: physics education at early-Victorian Cambridge, Edinburgh and Glasgow Universities', in *Wranglers and Physicists. Studies on Cambridge Physics in the Nineteenth Century*, ed. P. M. Harman (Manchester, 1985): 12–48, esp. 21–2; Maxwell's astronomy notes in ULC Add. MSS 7655, V, m/4.

12 See *LP*, **1**: 4 note (21); MS in ULC Add. MSS 7655, V, d/2.

13 Notes in ULC Add. MSS 7655, V, m/1–3; see *LP*, **1**: 75 note (4).

14 Wilson, 'The educational matrix': 22.

15 See *LP*, **1**: 6 note (26).

16 See *LP*, **1**: 149 note (63) and *Stokes–Kelvin Correspondence*, **1**: 46.

17 G. G. Stokes, 'On the theories of the internal friction of fluids in motion, and of the equilibrium and motion of elastic solids', *Transactions of the Cambridge Philosophical Society*, **8** (1845): 287–319, esp. 289–90, 312–13.

18 S. D. Poisson, 'Mémoire sur l'équilibre et le mouvement des corps élastiques', *Mémoires de l'Académie Royale des Sciences de l'Institut de France*, **8** (1829): 357–570, 623–7, on 361.

19 G. B. Airy, *Mathematical Tracts on the Lunar and Planetary Theories, the Figure of the Earth, Precession and Nutation, the Calculus of Variations, and the Undulatory Theory of Optics* (Cambridge, $_{2}$1831): iv; Thomson in *Electrostatics and Magnetism*: 15–37, 100–3; Stokes, 'On the theories of the internal friction of fluids': 290.

20 See Chapter III.1.

21 *Life of Maxwell*: 133.

22 J. D. Forbes to W. Whewell, 30 September 1850, 23 January 1851, 18 February 1851, 2 May 1852 in *LP*, **1**: 8–9.

23 *Isaac Newton's Philosophiae Naturalis Principia Mathematica. The Third Edition (1726) with Variant Readings*, ed. A. Koyré and I. B. Cohen, 2 vols. (Cambridge, 1972), **1**: 87–8; translation from *The Mathematical Papers of Isaac Newton*, ed. D. T. Whiteside, 8 vols. (Cambridge, 1967–81), **6**: 123.

24 Colin MacLaurin, *A Treatise of Fluxions*, 2 vols. (Edinburgh, 1742), **1**: 420.

25 *LP*, **1**: 8.

26 *LP*, **1**: 50 note (7).

27 John Leslie, *Elements of Geometry, Geometrical Analysis, and Plane Trigonometry* (Edinburgh, $_{2}$1811): vi, 2; Dugald Stewart, *Elements of the Philosophy of the Human Mind*, 3 vols. (Edinburgh, 1792–1827), **3**: 269; and see G. E. Davie, *The Democratic Intellect. Scotland and her Universities in the Nineteenth Century* (Edinburgh, $_{2}$1964): 105–200.

28 The comment by Babbage, Herschel and Peacock in their translation of S. F. Lacroix's text *Elementary Treatise on the Differential and Integral Calculus* (Cambridge, 1816): iii, and Peacock's note to the translation of the *Elementary Treatise*: 596, 604.

29 *The Cambridge University Calendar for the Year 1850* (Cambridge, 1850): 12. See

H. Becher, 'William Whewell and Cambridge mathematics', *Historical Studies in the Physical Sciences*, **11** (1980): 1–48; and I. Grattan-Guinness, 'Mathematics and mathematical physics from Cambridge, 1815–40', in *Wranglers and Physicists*, ed. Harman: 84–111.

30 W. Whewell, *The Doctrine of Limits* (Cambridge, 1838): 21.

31 Maxwell's notebook on 'Differential and Integral Calculus' in ULC Add. MSS 7655, V, m/7, f.9; Stokes' notes 'Differential Calculus N°. 1' in ULC Add. MSS 7656, PA2; and see *LP*, **1**: 10 note (44).

32 Becher, 'William Whewell and Cambridge mathematics'; Wilson, 'The educational matrix': 14–19.

33 Wilson, 'The educational matrix': 16.

34 ULC Add. MSS 7655, V, m/8–10; see *LP*, **1**: 11 note (48).

35 *Life of Maxwell*: 133.

36 *Cambridge Calendar for 1854*: 372–416.

37 Newton, *Principia*, **1**: 54; *Mathematical Papers*, ed. Whiteside, **6**: 97.

38 Newton, *Principia*, **1**: 54; *Mathematical Papers*, ed. Whiteside, **6**: 99.

39 *Leonhardi Euleri Opera Omnia*, ser. II, **5**, ed. J. O. Fleckenstein (Lausanne, 1957): 81, 89.

40 W. Whewell, *An Elementary Treatise on Mechanics* (Cambridge, $_7$1848): 132, 138.

41 Maxwell's notebook on 'Statics Dynamics' in ULC Add. MSS 7655, V, m/10, ff.58–60; and see Stokes' notes in ULC Add. MSS 7656, PA6 and Thomson's notes in ULC Add. MSS 7342, PA11. Compare W. Whewell, *A Treatise on Dynamics* (Cambridge, 1823): 4.

42 J. Clerk Maxwell, *Matter and Motion*, reprint edition (New York, n.d.): 48; compare *Natural Philosophy*: 163–81.

43 *LP*, **1**: 219 note (5).

44 David B. Wilson, *Kelvin and Stokes. A Comparative Study in Victorian Physics* (Bristol, 1987): 44.

45 W. Hopkins, *Remarks on Certain Proposed Regulations respecting the Studies of the University* (Cambridge, 1841): 10.

46 A contemporary recollection is cited in *Life of Maxwell*: 175n.

47 Wilson, 'The educational matrix': 19–26; Davie, *Democratic Intellect*: 178–89; R. G. Olson, *Scottish Philosophy and British Physics 1750–1880* (Princeton, 1975): esp. 287–231 on Maxwell.

48 J. F. W. Herschel, *A Preliminary Discourse on the Study of Natural Philosophy* (London, 1830): 96.

49 *Life of Maxwell*: 108.

50 Thomas Reid, *Essays on the Intellectual Powers of Man*, ed. A. D. Woozley (London, 1941): 383.

51 Dugald Stewart, *Philosophical Essays* (Edinburgh, $_3$1818): 153; Stewart, *Philosophy of the Human Mind*, **2**: 193; and see *Lectures on Metaphysics by Sir William Hamilton*, ed. H. L. Mansel and J. Veitch, 2 vols. (Edinburgh/London, 1859), **2**: 112.

52 Hamilton, *Lectures on Metaphysics*, **1**: 160, **2**: 113–14.

53 Hamilton, *Lectures on Metaphysics*, **2**: 279.

54 MacLaurin, *Fluxions*, **1**: 51–2; Stewart, *Philosophy of the Human Mind*, **2**: 386.

55 Hamilton, *Lectures on Metaphysics*, **1**: 160.
56 William Whewell, *The Philosophy of the Inductive Sciences, Founded upon their History*, 2 vols. (London, $_2$1847), **1**: 16, 18.
57 W. Whewell to J. D. Forbes, 10 March 1855 in *LP*, **1**: 12, and Maxwell to L. Campbell, 21 November 1865 in *LP*, **2**: 228.
58 C. A. Bristed, *Five Years in an English University*, 2 vols. (New York, 1852), **1**: 386–96.
59 Whewell, *Philosophy*, **2**: 650.
60 Whewell, *Philosophy*, **1**: 59, 66, 80, 90. See R. E. Butts, 'Necessary truth in Whewell's theory of science', *American Philosophical Quarterly*, **2** (1965): 1–21.
61 Whewell, *Philosophy*, **1**: 215–45, **2**: 573–94. See P. M. Harman, *Metaphysics and Natural Philosophy* (Brighton, 1982): 56–80.
62 Maxwell's notebook 'Statics Dynamics' in ULC Add. MSS 7655, V, m/10, ff.62–3; see the notes by Stokes in ULC Add. MSS 7656, PA6 and by Thomson in ULC Add. MSS 7342, PA11, ff.10–11.
63 Maxwell, *Matter and Motion*: 29.
64 Whewell, *Philosophy*, **1**: x.
65 Hamilton, *Lectures on Metaphysics*, **2**: 260.

Chapter III Edinburgh physics and Cambridge mathematics

1 Isaac Newton, *Opticks*, Dover reprint based on the fourth edition of 1730 (New York, 1952): 155–8. See Alan E. Shapiro, 'The evolving structure of Newton's theory of white light and color', *Isis*, **71** (1980): 211–35, esp. 234–5.
2 *The Correspondence of Isaac Newton*, **1**, ed. H. W. Turnbull (Cambridge, 1959): 291.
3 Newton, *Opticks*: 133. See Alan E. Shapiro, 'Artists' colors and Newton's colors', *Isis*, **85** (1994): 600–30.
4 J. D. Forbes, 'Hints towards a classification of colours', *Phil. Mag.*, ser. 3, **34** (1849): 161–78, on 162, 165, 168–9.
5 R. L. Kremer, 'Innovation through synthesis: Helmholtz and colour research', in *Hermann von Helmholtz and the Foundations of Nineteenth-Century Science*, ed. D. Cahan (Berkeley, Cal., 1993): 205–58, esp. 222–3, 232.
6 H. Helmholtz, 'On the theory of compound colours', *Phil. Mag.*, ser. 4, **4** (1852): 519–34, on 526, 528.
7 J. D. Forbes' notebook in *LP*, **1**: 301 note (5); Forbes to Maxwell, 4 May 1855 in *LP*, **1**: 300 note (2).
8 Forbes, 'Hints towards a classification of colours': 172.
9 See *LP*, **1**: 245–6 note (12).
10 H. Grassmann, 'On the theory of compound colours', *Phil. Mag.*, ser. 4, **7** (1854): 254–64, esp. 255. See *LP*, **1**: 268 note (5) and P. D. Sherman, *Colour Vision in the Nineteenth Century* (Bristol, 1981): 93–116.
11 See *LP*, **1**: 269 note (6).
12 W. Thomson to G. G. Stokes, 28 January 1856 and Stokes to Thomson, 4 February 1856 in *LP*, **1**: 326 note (7) and *Stokes–Kelvin Correspondence*, **1**: 209–10.
13 G. G. Stokes to Maxwell, 7 November 1857 in *LP*, **1**: 568 note (2).

14 David Brewster, 'On a new analysis of solar light', *Transactions of the Royal Society of Edinburgh*, **12** (1834): 123–36. See *LP*, **1**: 571 note (15).
15 H. Helmholtz, 'On Sir David Brewster's new analysis of solar light', *Phil. Mag.*, ser. 4, **4** (1852): 401–16.
16 *LP*, **1**: 647 note (6).
17 *LP*, **1**: 645 note (2), 647 note (6).
18 J. Challis to W. Thomson, 28 February 1855 in S. G. Brush, C. W. F. Everitt and E. Garber, eds., *Maxwell on Saturn's Rings* (Cambridge, Mass., 1983): 6–7.
19 'Inner ring of Saturn', *Monthly Notices of the Royal Astronomical Society*, **11** (1851): 20–7.
20 Edward S. Holden, *Memorials of William Cranch Bond . . . and of his Son George Phillips Bond* (San Francisco/New York, 1897): 102–3.
21 Otto Struve, 'Sur les dimensions des anneaux de Saturne', *Mémoires de l'Académie Impériale des Sciences de Saint-Petersbourg*, ser. 6, **5** (1853): 439–75, on 444, 473; abstract in *Monthly Notices of the Royal Astronomical Society*, **13** (1852): 22–4.
22 Holden, *Memorials*: 112–17.
23 J. Challis to W. Thomson, 14 March 1855 in Brush, Everitt and Garber, *Saturn's Rings*: 8–9.
24 Challis' draft with Thomson's emendations is reproduced in P. M. Harman, 'Maxwell and Saturn's rings: problems of stability and calculability', in *The Investigation of Difficult Things. Essays on Newton and the History of the Exact Sciences in Honour of D. T. Whiteside*, ed. P. M. Harman and Alan E. Shapiro (Cambridge, 1992): 477–502, on 501–2.
25 J. Challis to W. Thomson, 14 March 1855 in Brush, Everitt, and Garber, *Saturn's Rings*: 8–9.
26 *The Cambridge University Calendar for the Year 1854* (Cambridge, 1854): 413.
27 P. S. de Laplace, *Traité de Mécanique Céleste*, 5 vols. (Paris, 1799–1825), **2**: 155–66.
28 Joseph Plateau, 'On the phenomena presented by a free liquid mass withdrawn from the action of gravity', in *Scientific Memoirs*, ed. R. Taylor, **4** (London, 1846): 16–43, on 27–8, 35–6.
29 Laplace, *Mécanique Céleste*, **5**: 291.
30 *LP*, **1**: 438–9 notes (2) and (4).
31 G. B. Airy, 'On the stability of the motion of Saturn's rings', *Monthly Notices of the Royal Astronomical Society*, **19** (1859): 297–304, on 300.
32 Otto Mayr, 'Maxwell and the origins of cybernetics', *Isis*, **62** (1971): 425–44, esp. 428–9; Harman, 'Maxwell and Saturn's rings': 488, 495–6; and see Maxwell to W. Thomson, 11 September 1863 in *LP*, **2**: 112–16.
33 R. B. Hayward, 'On a direct method of estimating velocities, accelerations and all similar quantities with respect to axes, moveable in any manner in space, with applications', *Transactions of the Cambridge Philosophical Society*, **10** (1856): 1–20, esp. 7–10.
34 P. M. Harman, *Energy, Force, and Matter. The Conceptual Development of Nineteenth-century Physics* (Cambridge, 1982): 58–64; W. J. M. Rankine, 'On the general law of the conservation of energy', *Phil. Mag.*, ser. 4, **5** (1853): 106–17, on 106.

35 *LP*, **1**: 473 note (w).
36 Wilhelm Weber, 'On the measurement of electrodynamic forces', in *Scientific Memoirs*, ed. R. Taylor, **5** (London, 1852): 489–529. See C. Jungnickel and R. McCormmach, *Intellectual Mastery of Nature. Theoretical Physics from Ohm to Einstein*, 2 vols. (Chicago/London, 1986), **1**: 65–77, 138–44.
37 Bruce J. Hunt, 'The Ohm is where the art is: British telegraph engineers and the development of electrical standards', *Osiris*, 2nd ser., **9** (1994): 48–63.
38 'Report of the Committee . . . on standards of electrical resistance', *Report of the Thirty-third Meeting of the British Association for the Advancement of Science; held . . . in August and September 1863* (London, 1864): 111–76, on 111–20. See *LP*, **2**: 88–9 note (2) and C. W. Smith and M. N. Wise, *Energy and Empire. A Biographical Study of Lord Kelvin* (Cambridge, 1989): 687–94.
39 Simon Schaffer, 'Accurate measurement is an English science', in *The Values of Precision*, ed. M. N. Wise (Princeton, 1995): 135–72.
40 J. Clerk Maxwell and Fleeming Jenkin, 'On the elementary relations between electrical measurements' included in the 'Report of the Committee . . . on standards of electrical resistance': 130–63, reprinted with corrections in *Phil. Mag.*, ser. 4, **29** (1865): 436–60, 507–25.
41 Daniel M. Siegel, *Innovation in Maxwell's Electromagnetic Theory* (Cambridge, 1991): 136.
42 Léon Foucault, 'Détermination expérimentale de la vitesse de la lumière: parallaxe du soleil', *Comptes Rendus*, **55** (1862): 501–3. See Schaffer, 'Accurate measurement is an English science': 146–8.
43 See C. Hockin to Maxwell, 15 May 1868 in *LP*, **2**: 376 note (8).
44 Siegel, *Innovation in Maxwell's Electromagnetic Theory*: 156.
45 Dugald M'Kichan, 'Determination of the number of electrostatic units in the electromagnetic unit made in the Physical Laboratory of Glasgow University', *Philosophical Transactions of the Royal Society*, **163** (1873): 409–27.
46 See *Cambridge University Reporter* (19 October 1870 to 15 February 1871) in *LP*, **2**: 33–4 esp. notes (104) to (106).
47 G. G. Stokes to Maxwell, 18 February 1871 in *LP*, **2**: 612 note (4).
48 *LP*, **2**: 35.
49 J. W. Strutt to Maxwell, 14 February 1871 in *LP*, **2**: 611–12 note (3).
50 Maxwell to Fleeming Jenkin, 18 November 1874, ULC Add. MSS 7655, II/242; and see Maxwell to H. A. Rowland, 9 July 1874, Rowland Papers MS 6, Milton S. Eisenhower Library, The Johns Hopkins University, Baltimore.

Chapter IV Physical and geometrical analogy

1 *Electrostatics and Magnetism*: 347.
2 P. M. Heimann, 'Maxwell and the modes of consistent representation', *Archive for History of Exact Sciences*, **6** (1970): 171–213.
3 *Electricity*, **1**: 8, 16, 32n, 64, 66–9.
4 *Electricity*, **1**: 362, 383, 386, 393, 409–10, 531n.

5 *Electricity*, **3**: 30, 414.
6 *Electricity*, **3**: 418–22, plate IV fig. 1.
7 *Electricity*, **3**: 349. See M. N. Wise, 'The mutual embrace of electricity and magnetism', *Science*, **203** (1979): 1310–18.
8 *Electrostatics and Magnetism*: 4–5. See Ole Knudsen, 'Mathematics and physical reality in William Thomson's electromagnetic theory', in *Wranglers and Physicists*, ed. P. M. Harman (Manchester, 1985): 149–79.
9 On the term 'potential' see *LP*, **1**: 258 note (13).
10 *Electrostatics and Magnetism*: 28–9.
11 C. W. Smith and M. N. Wise, *Energy and Empire. A Biographical Study of Lord Kelvin* (Cambridge, 1989): 212–19; Knudsen, 'William Thomson's electromagnetic theory': 150–7.
12 *Electrostatics and Magnetism*: 37.
13 *Electrostatics and Magnetism*: 98–9; see Smith and Wise, *Energy and Empire*: 238–40.
14 *Math. & Phys. Papers*, **1**: 76–80.
15 See esp. W. Thomson, 'Notes on hydrodynamics. I. On the equation of continuity', *Cambridge and Dublin Mathematical Journal*, **2** (1847): 282–6. Maxwell transcribed the first part of this paper almost verbatim in his undergraduate notes on 'Hydrodynamics', ULC Add. MSS 7655, V, m/8.
16 *Electrostatics and Magnetism*: 378–404. On Ampère's term 'solénoide électro-dynamique' see *LP*, **1**: 323 note (29).
17 *Electrostatics and Magnetism*: 388n.
18 Smith and Wise, *Energy and Empire*: 203–81; M. N. Wise, 'The flow analogy to electricity and magnetism', *Archive for History of Exact Sciences*, **25** (1981): 19–70.
19 Wise, 'The mutual embrace of electricity and magnetism': 1310–18.
20 Wise, 'The mutual embrace of electricity and magnetism': 1314–15.
21 *Stokes–Kelvin Correspondence*, **1**: 97; *The Cambridge University Calendar for the Year 1854* (Cambridge, 1854): 415; and see *LP*, **1**: 257–8 note (12).

Chapter V Models and mechanisms

1 Rudolf Clausius, 'On the mean length of the paths described by the separate molecules of gaseous bodies on the occurrence of molecular motion', *Phil. Mag.*, ser. 4, **17** (1859): 81–91.
2 Peter Achinstein, *Particles and Waves. Historical Essays in the Philosophy of Science* (New York/Oxford 1991): 207–32.
3 R. Clausius, 'On the nature of the motion which we call heat', *Phil. Mag.*, ser. 4, **14** (1857): 108–27, on 108.
4 P. M. Harman, *Energy, Force, and Matter. The Conceptual Development of Nineteenth-century Physics* (Cambridge, 1982): 52–8.
5 *Natural Philosophy*: vi.
6 Clausius, 'On the mean length of the paths': 83–4.
7 On the development of Maxwell's kinetic theory of gases see Elizabeth Garber, Stephen G. Brush, and C. W. F. Everitt, eds., *Maxwell on Molecules and Gases* (Cambridge,

Mass., 1986): 1–47, and T. M. Porter, 'A statistical survey of gases: Maxwell's social physics', *Historical Studies in the Physical Sciences*, **12** (1981): 77–116.

8 See Maxwell to G. B. Airy, 16 October 1872 in *LP*, **2**: 758.

9 He used incorrect values in this latter calculation: see *SP*, **1**: 405n and *LP*, **1**: 660 note (8).

10 Daniel M. Siegel, *Innovation in Maxwell's Electromagnetic Theory. Molecular Vortices, Displacement Current, and Light* (Cambridge, 1991): 37–9.

11 William Thomson, 'Dynamical illustrations of the magnetic and the heliocoidal rotatory effects of transparent bodies on polarized light', *Phil. Mag.*, ser. 4, **13** (1857): 198–204, on 200.

12 W. Thomson to G. G. Stokes, 23 May 1857 and 23 December 1857; see *LP*, **1**: 560–1 notes (2) and (5) and *Stokes–Kelvin Correspondence*, **1**: 224, 229.

13 In the museum of the Cavendish Laboratory, Cambridge: see plate X in *LP*, **1**: facing p. 688.

14 On subsequent discussion of electromechanical effects see Everitt, *James Clerk Maxwell*: 107–8 and *LP*, **1**: 688 note (20).

15 Siegel, *Innovation in Maxwell's Electromagnetic Theory*: 65–73.

16 Siegel, *Innovation in Maxwell's Electromagnetic Theory*: 73–83.

17 *Electricity*, **1**: 409.

18 Siegel, *Innovation in Maxwell's Electromagnetic Theory*: 85–119, esp. 105–12.

19 Joan Bromberg, 'Maxwell's displacement current and his theory of light', *Archive for History of Exact Sciences*, **4** (1967): 218–34, on 219.

20 Siegel, *Innovation in Maxwell's Electromagnetic Theory*: 120–43.

21 See *LP*, **1**: 687 note (17).

22 For Maxwell's further comments on Verdet's work see his letters to P. G. Tait of 23 December 1867 (*LP*, **2**: 336) and to W. Thomson of 18 July 1868 (*LP*, **2**: 405–6). See also Ole Knudsen, 'The Faraday effect and physical theory, 1845–1873', *Archive for History of Exact Sciences*, **15** (1976): 235–81, esp. 255–61.

23 Siegel, *Innovation in Maxwell's Electromagnetic Theory*: 29–55.

Chapter VI Dynamical and statistical explanation

1 Thomson's letter is not extant.

2 W. J. M. Rankine, 'On the general law of the conservation of energy', *Phil. Mag.*, ser. 4, **5** (1853): 106–17.

3 P. M. Harman, 'Mathematics and reality in Maxwell's dynamical physics', in *Kelvin's Baltimore Lectures and Modern Theoretical Physics*, ed. R. Kargon and P. Achinstein (Cambridge, Mass., 1987): 267–97; and see S. D. Poisson, 'Mémoire sur l'équilibre et le mouvement des corps élastiques', *Mémoires de l'Académie Royale des Sciences de l'Institut de France*, **8** (1829): 357–570, 623–7, on 361.

4 Maxwell was commenting on Tait's remarks in a draft of his *Sketch of Thermodynamics* (Edinburgh, 1868); see *LP*, **2**: 335 note (2), 337 note (14) and Chapter VI.3.

5 Daniel M. Siegel, *Innovation in Maxwell's Electromagnetic Theory* (Cambridge, 1991): 146–50.

6 Siegel, *Innovation in Maxwell's Electromagnetic Theory*: 150–2; Jed Z. Buchwald, *From Maxwell to Microphysics. Aspects of Electromagnetic Theory in the Last Quarter of the Nineteenth Century* (Chicago/London, 1985): 20–40.
7 *Math. & Phys. Papers*, **1**: 107–12.
8 *Math. & Phys. Papers*, **4**: 459; C. G. Knott, *Life and Scientific Work of Peter Guthrie Tait* (Cambridge, 1911): 182–3.
9 *Natural Philosophy*: 217–25; and see *LP*, **2**: 763 note (3) and C. W. Smith and M. N. Wise, *Energy and Empire. A Biographical Study of Lord Kelvin* (Cambridge, 1989): 390–5.
10 C. G. Gauss, *Werke*, **5** (Göttingen, 1867): 629.
11 [John Herschel,] 'Quetelet on probabilities', *Edinburgh Review*, **92** (July 1850): 1–57. On the date of Maxwell's letter to Campbell and the context see *LP*, **1**: 193 note (1), 194 note (4), 197 note (11).
12 *LP*, **1**: 247 and *Life of Maxwell*: 113. See *LP*, **1**: 197 note (11) and T. M. Porter, *The Rise of Statistical Thinking 1820–1900* (Princeton, 1986): 40–55, 60–5, 78–83, 110–28.
13 Porter, *Rise of Statistical Thinking*: 113, 116.
14 J. F. W. Herschel, *Essays from the Edinburgh and Quarterly Reviews, with Addresses and other Pieces* (London, 1857): 365–465, esp. 398–400. See S. G. Brush, 'Foundations of statistical mechanics 1845–1915', *Archive for History of Exact Sciences*, **4** (1967): 145–83, esp. 151–2, and Porter, *Rise of Statistical Thinking*: 118–25.
15 J. Clerk Maxwell, *Theory of Heat* (London, 1871): 287–9, 308–9.
16 See H. G. van Leeuwen, *The Problem of Certainty in English Thought 1630–1690* (The Hague, $_2$1970): 143–52.
17 J. Clerk Maxwell, 'The kinetic theory of gases', *Nature*, **16** (1877): 242–6, on 242.
18 G. Gigerenzer *et al.*, *The Empire of Chance* (Cambridge, 1989): 185–7.
19 Royal Society, *Referees' Reports*, **8**: 188.
20 P. G. Tait to Maxwell, 6 December 1867 in *LP*, **2**: 328 note (2). On Tait's draft chapters of his *Sketch of Thermodynamics* see *LP*, **2**: 335 note (2).
21 *Math. & Phys. Papers*, **1**: 118–19n.
22 *Math. & Phys. Papers*, **1**: 232.
23 *Math. & Phys. Papers*, **1**: 179, 511–14.
24 W. Thomson to G. G. Stokes, 13 October 1866 in *Stokes–Kelvin Correspondence*, **1**: 329 and *LP*, **2**: 292 note (8).
25 See *LP*, **2**: 853–5, 937–9; *SP*, **2**: 740–1.
26 MS 'Concerning Demons', ULC Add. MSS 7655, V, i/11a; and in Knott, *Life and Scientific Work of Peter Guthrie Tait*: 214–15. For a full and illuminating analysis see Martin J. Klein, 'Maxwell, his demon, and the second law of thermodynamics', *American Scientist*, **58** (1970): 84–97.
27 On Boltzmann's explication of this issue see Klein, 'Maxwell, his demon, and the second law of thermodynamics': 91–2.
28 *Math. & Phys. Papers*, **5**: 12n.
29 MS cited in note (26).
30 Maxwell, *Theory of Heat*: 153–4.
31 Maxwell, *Theory of Heat*: 308–9.

32 Martin J. Klein, 'Gibbs on Clausius', *Historical Studies in the Physical Sciences*, **1** (1969): 127–49; E. E. Daub, 'Entropy and dissipation', *ibid.*, **2** (1970): 321–54. See Maxwell to P. G. Tait, 1 December 1873 in *LP*, **2**: 945–6 for his clarification of his confusion over the concept of entropy.
33 Maxwell's thermodynamic *nom de plume* expressed the second law of thermodynamics, taking its form, $dp/dt = J\,C\,M$, from Tait's expression for the law in his *Sketch of Thermodynamics*: 91. See *LP*, **2**: 543–4 note (17), and (for a full analysis) Klein, 'Maxwell, his demon, and the second law of thermodynamics': 94–5.
34 Maxwell to P. G. Tait, 13 October 1876, ULC Add. MSS 7655, I, b/83.
35 Klein, 'Maxwell, his demon, and the second law of thermodynamics': 87–9.
36 See *LP*, **2**: 361 note (12).

Chapter VII Geometry and physics

1 P. G. Tait to Maxwell, 13 December 1867 in *LP*, **2**: 334 note (22). Tait's mode of writing the operator ∇ here has been changed for consistency.
2 *The Mathematical Papers of Sir William Rowan Hamilton. Vol. III. Algebra*, ed. H. Halberstam and R. E. Ingram (Cambridge, 1967): 4–7. See T. L. Hankins, *Sir William Rowan Hamilton* (Baltimore/London, 1980): 268–75.
3 *Mathematical Papers of Hamilton*, **3**: 355–9.
4 *Mathematical Papers of Hamilton*, **3**: 117, 117n, 145.
5 P. G. Tait, *An Elementary Treatise on Quaternions* (Oxford, 1867): 221, 267.
6 *Mathematical Papers of Hamilton*, **3**: 263, 376–7; W. R. Hamilton, *Lectures on Quaternions* (Dublin, 1853): 610–11.
7 P. G. Tait, 'On Green's and other allied theorems', *Transactions of the Royal Society of Edinburgh*, **26** (1870): 69–84.
8 See *LP*, **2**: 577 notes (2) and (3).
9 In the second edition of *A Treatise on Electricity and Magnetism*, ed. W. D. Niven, 2 vols. (Oxford, 1881), **1**: 12 (§12) Maxwell uses the terms 'Intensities' and 'Fluxes'.
10 *Natural Philosophy*. 124.
11 *Stokes–Kelvin Correspondence*, **1**: 97.
12 Tait, 'On Green's and other allied theorems': 75; Tait to Maxwell, 5 April 1871 in *LP*, **2**: 634 note (5).
13 J. B. Listing, 'Vorstudien zur Topologie', in *Göttinger Studien. 1847. Erste Abtheilung: Mathematische und naturwissenschaftliche Abhandlungen* (Göttingen, 1847): 811–75.
14 C. G. Gauss, *Werke*, **5** (Göttingen, 1867): 605.
15 Hermann Helmholtz, 'Über Integrale der hydrodynamischen Gleichungen, welche den Wirbelbewegungen entsprechen', *Journal für die reine und angewandte Mathematik*, **55** (1858): 22–55; trans. by P. G. Tait as 'On the integrals of the hydrodynamical equations, which express vortex motion', *Phil. Mag.*, ser. 4, **33** (1867): 485–512.
16 B. Riemann, 'Lehrsätze aus der analysis situs für die Theorie der Integrale von zweigliedrigen vollständigen Differentialen', *Journal für die reine und angewandte Mathematik*, **54** (1857): 105–10.
17 See *Proceeding of the Royal Society of Edinburgh*, **6** (1867): 167 and *Math. & Phys. Papers*, **4**: 13–66.

18 J. B. Listing, 'Der Census räumlicher Complexe oder Verallgemeinerung des Euler'schen Satzes von den Polyëdern', *Abhandlungen der Math. Classe der Königlichen Gesellschaft der Wissenschaften zu Göttingen*, **10** (1861): 97–182.

19 See *LP*, **2**: 466 note (2).

20 Hamilton, *Lectures on Quaternions*: 59; *Natural Philosophy*: 173.

21 Listing, 'Vorstudien zur Topologie': 830, 838–50.

22 Hankins, *Hamilton*: 326–33.

Chapter VIII Physical reality: ether and matter

1 William Whewell, *The Philosophy of the Inductive Sciences, Founded upon their History*, 2 vols. (London, $_{2}$1847), **2**: 65–74.

2 G. F. FitzGerald, 'On the electromagnetic theory of the reflection and refraction of light', *Philosophical Transactions of the Royal Society*, **171** (1880): 691–711, on 691; *Memoir and Scientific Correspondence of Sir George Gabriel Stokes, Bart.*, ed. J. Larmor, 2 vols. (Cambridge, 1907): **2**: 42.

3 H. Fizeau, 'On the effect of the motion of a body upon the velocity with which it is traversed by light", *Phil. Mag.*, ser. 4, **19** (1860): 245–60.

4 A summary of this work is appended to the translation of Fizeau's paper cited in note (3): see *Phil. Mag.*, ser. 4, **19** (1860): 258–60.

5 See the 'Lettre de M. Fresnel à M. Arago', *Annales de Chimie et de Physique*, **9** (1818): 57–66.

6 G. G. Stokes, 'On Fresnel's theory of the aberration of light', *Phil. Mag.*, ser. 3, **28** (1846): 76–81.

7 W. Huggins, 'Further observations on the spectra of some of the stars and nebulae . . .', *Phil. Trans.*, **158** (1868): 529–64, on 532–5.

8 J. Clerk Maxwell, 'On a possible mode of detecting a motion of the solar system through the luminiferous ether', *Proceedings of the Royal Society*, **30** (1880): 108–10 and *Nature*, **21** (1880): 314–15.

9 Jules Jamin, 'Note sure la théorie de la réflection et de la réfractioń, *Annales de Chimie et de Physique*, ser. 3, **59** (1860): 413–26.

10 James MacCullagh, 'On the laws of crystalline reflexion and refraction', *Transactions of the Royal Irish Academy*, **18** (1837); 31–74. On MacCullagh and Neumann's papers see *LP*, **2**: 182–4 notes (3) to (9), 186–7 notes (4) to (7).

11 Bruce J. Hunt, *The Maxwellians* (Ithaca/London, 1991): 28–30; and Thomas K. Simpson, 'Maxwell and the direct experimental test of his electromagnetic theory', *Isis*, **57** (1966): 411–32.

12 Jed Z. Buchwald, *The Creation of Scientific Effects. Heinrich Hertz and Electric Waves* (Chicago/London, 1994).

13 *Correspondence of Stokes*, **2**: 42.

14 Bernhard Riemann, 'Ein Beitrag zur Electrodynamik', *Annalen der Physik und Chemie*, **131** (1867): 237–43; (trans.) 'A contribution to electrodynamics', *Phil. Mag.*, ser. 4, **34** (1867): 368–72.

15 Carl Neumann, 'Resultate einer Untersuchung über die Principien der Elektrodynamik', *Nachrichten von der Königl. Gesellschaft der Wissenschaften der Georg-August-Universität zu Göttingen* (1868): 223–35.

16 Hermann Helmholtz, *Über die Erhaltung der Kraft, eine physikalische Abhandlung* (Berlin, 1847): 63; (trans.) 'On the conservation of force', in *Scientific Memoirs, Natural Philosophy*, ed. J. Tyndall and W. Francis (London, 1853): 114–62, esp. 156.

17 *Electricity*, **3**: 532n, 571.

18 *Electricity*, **3**: 574.

19 Michael Faraday, 'On the conservation of force', *Phil. Mag.*, ser. 4, **13** (1857): 225–39, on 228–9.

20 Helmholtz, 'On the conservation of force': 122; W. J. M. Rankine, 'On the general law of the transformation of energy', *Phil. Mag.*, ser. 4, **5** (1853): 106–17, esp. 106.

21 Rudolf Clausius, 'On the conduction of heat by gases', *Phil. Mag.*, ser. 4, **23** (1862): 417–35, 512–34, on 425n, 527–8n. See *LP*, **2**: 73 note (99), 83 note (39).

22 See also *LP*, **2**: 96 note (11).

23 E. Garber, S. G. Brush, and C. W. F. Everitt, eds., *Maxwell on Molecules and Gases* (Cambridge, Mass., 1986): 40, 60.

24 William Crookes, 'On attraction and repulsion resulting from radiation', *Philosophical Transactions of the Royal Society*, **164** (1874): 501–27. The paper was read to the Royal Society on 11 December 1873: see *Proceedings of the Royal Society*, **22** (1873): 37–41.

25 Royal Society, *Referees' Reports*, **7**: 295.

26 Maxwell to P. G. Tait [late April 1874], ULC Add. MSS 7655, I, b/65.

27 On the circumstances of Maxwell's paper see S. G. Brush and C. W. F. Everitt, 'Maxwell, Osborne Reynolds, and the radiometer', *Historical Studies in the Physical Sciences*, **1** (1969): 105–25.

28 Thomas Andrews, 'On the continuity of the gaseous and liquid states of matter', *Philosophical Transactions of the Royal Society*, **159** (1869): 575–89, esp. 588–9.

29 See Kostas Gavroglu, 'The reaction of the British physicists and chemists to van der Waals' early work and to the law of corresponding states', *Historical Studies in the Physical Sciences*, **20** (1990): 199–237, esp. 218–19 for comment on Maxwell's criticisms of van der Waals.

30 Royal Society, *Referees' Reports*, **7**: 434.

31 *Isaac Newton's Philosophiae Naturalis Principia Mathematica. The Third Edition (1726) with Variant Readings*, ed. A. Koyré and I. B. Cohen, 2 vols. (Cambridge, 1972), **2**: 552–4. See. P. M. Harman, *Metaphysics and Natural Philosophy* (Brighton, 1982): 18–22.

32 W. Whewell, *The Philosophy of the Inductive Sciences, Founded upon their History*, 2 vols. (London, $_{2}$1847), **1**: 432, **2**: 289.

33 MS on 'Dimensions of Physical Quantities', ULC Add. MSS 7655, V, h/4.

34 J. Clerk Maxwell, 'The kinetic theory of gases', *Nature*, **16** (1877): 242–6, on 245; see Chapter VI.2.

35 J. F. W. Herschel, *A Preliminary Discourse on the Study of Natural Philosophy* (London, 1830): 38.

Notes to Chapter IX

Chapter IX Physics and metaphysics

1 Maxwell to G. Chrystal, 22 July 1874, ULC Add. MSS 8375/1.
2 J. Clerk Maxwell, *Matter and Motion*, reprint edition (New York, n.d.): 29.
3 [J. Fletcher Moulton,] 'Herbert Spencer', *The British Quarterly Review*, **58** (October 1873): 472–504; Herbert Spencer, 'Replies to criticisms', *Fortnightly Review*, **14** (November and December 1873): 581–95, 715–39, esp. 736–7; *Natural Philosophy*: 178.
4 P. G. Tait, *Sketch of Thermodynamics* (Edinburgh, 1868): 1; Tait, 'Herbert Spencer versus Thomson and Tait', *Nature*, **9** (26 March 1874): 402–3; H. Spencer, 'Prof. Tait and Mr. Spencer', *ibid.* (2 April 1874): 420–1.
5 On the context see *LP*, **2**: 335 note (3).
6 *Natural Philosophy*: 161.
7 *Lezioni Accademiche D'Evangelista Torricelli* (Florence, 1715): 25.
8 See *LP*, **2**: 365 note (12).
9 See *LP*, **2**: 395 note (29).
10 W. Thomson and P. G. Tait, *Treatise on Natural Philosophy, Vol. 1, Part 1*, new edition (Cambridge, 1879): 219, as in *Natural Philosophy*: 161.
11 *Natural Philosophy*: 164, *Natural Philosophy* (new edition): 222.
12 *Isaac Newton's Philosophiae Naturalis Principia Mathematica. The Third Edition (1726) with Variant Readings*, ed. A. Koyré and I. B. Cohen, 2 vols. (Cambridge, 1972), **1**: 40; translation from *The Mathematical Papers of Isaac Newton*, ed. D. T. Whiteside, 8 vols. (Cambridge, 1967–81), **6**: 93.
13 *Electricity*, **2**: 290–2, **3**: 447.
14 P. M. Harman, *Metaphysics and Natural Philosophy* (Brighton, 1982): 81–104.
15 B. Riemann, 'On the hypotheses which lie at the bases of geometry', *Nature*, **8** (1873): 14–17, 36–7.
16 W. K. Clifford, 'On the space-theory of matter', *Proceedings of the Cambridge Philosophical Society*, **2** (1870): 157–8.
17 Maxwell to P. G. Tait, 11 November 1874, ULC Add. MSS 7655, I, b/72: printed in facsimile in P. M. Harman, *Energy, Force, and Matter. The Conceptual Development of Nineteenth-century Physics* (Cambridge, 1982): 96.
18 Horace, *Satires*, II. 7. 86.
19 Maxwell, *Matter and Motion*: 12.
20 Maxwell to P. G. Tait, 19 March 1875, ULC Add. MSS 7655, I, b/76.
21 See also *LP*, **2**: 363 note (8).
22 W. Thomson, 'On the ultramundane corpuscles of Le Sage', *Proceedings of the Royal Society of Edinburgh*, **7** (1871): 577–89.
23 John Tyndall, *Fragments of Science: A Series of Detatched Essays, Addresses and Reviews*, 2 vols. (London, $_6$1879), **2**: 75–90.
24 Tyndall, *Fragments of Science*, **2**: 137–203 on 193–4. See Ruth Barton, 'John Tyndall, pantheist: a rereading of the Belfast address', *Osiris*, 2nd ser., **3** (1987): 111–34.
25 Frank M. Turner, *Contesting Cultural Authority. Essays in Victorian Intellectual Life* (Cambridge, 1993): 131–50, 171–200.
26 Tyndall, *Fragments of Science*, **2**: 182–3.

27 Tyndall, *Fragments of Science*, **2**: 199.

28 *Life of Maxwell*: 639.

29 [Balfour Stewart and Peter Guthrie Tait,] *The Unseen Universe: or Physical Speculations on a Future State* (London, $_2$1875): xii, 158–9. 164, 172. See P. M. Heimann, 'The *Unseen Universe*: physics and the philosophy of nature in Victorian Britain', *The British Journal for the History of Science*, **6** (1972): 73–9.

30 *Life of Maxwell*: 394, 405.

31 See *LP*, **2**: 2.

32 Hermann von Helmholtz, *Wissenschaftliche Abhandlungen*, 3 vols. (Leipzig, 1882–95), **3**: 579.

33 [Fleeming Jenkin,] 'The atomic theory of Lucretius', *North British Review*, **48** (1868): 211–42. See Turner, *Contesting Cultural Authority*: 262–83; and, more broadly, see T. M. Porter, *The Rise of Statistical Thinking 1820–1900* (Princeton, 1986): 194–208, and C. W. Smith and M. N. Wise, *Energy and Empire. A Biographical Study of Lord Kelvin* (Cambridge, 1989): 628–33.

34 H. A. J. Munro's translation, in his edition of *Titi Lucreti Cari De Rerum Natura*, 2 vols. (Cambridge, $_2$1866), **1**: 93–4, is cited here. See *LP*, **2**: 655 note (5).

35 Maxwell to F. Galton, 26 February 1879, in extract in Porter, *Rise of Statistical Thinking*: 205–6; and see Joseph Boussinesq, *Conciliation du Veritable Déterminisme Mécanique avec l'Existence de la Vie et de la Liberté Morale* (Paris, 1878): 50.

Index

Bold figures refer to text numbers in *LP*, **1** and **2**.

Index

Index

Index

Index

Index

Index

Index

Index